An Investigation of UV Disinfection and Repair in *Cryptosporidium parvum*

The mission of the Awwa Research Foundation (AwwaRF) is to advance the science of water to improve the quality of life. Funded primarily through annual subscription payments from over 1,000 utilities, consulting firms, and manufacturers in North America and abroad, AwwaRF sponsors research on all aspects of drinking water, including supply and resources, treatment, monitoring and analysis, distribution, management, and health effects.

From its headquarters in Denver, Colorado, the AwwaRF staff directs and supports the efforts of over 800 volunteers, who are the heart of the research program. These volunteers, serving on various boards and committees, use their expertise to select and monitor research studies to benefit the entire drinking water community.

Research findings are disseminated through a number of technology transfer activities, including research reports, conferences, videotape summaries, and periodicals.

An Investigation of UV Disinfection and Repair in *Cryptosporidium parvum*

Prepared by:
Paul A. Rochelle and **Alexander A. Mofidi**
Metropolitan Water District of Southern California
La Verne, CA 91750

Marilyn M. Marshall
University of Arizona
Tucson, AZ 85721

Steve J. Upton, **Beth Montelone**, and **Keith Woods**
Kansas State University
Manhattan, KS 66506

and

George Di Giovanni
Texas Agricultural Experiment Station
Texas A&M University System
El Paso, TX 79927

Sponsored by:
Awwa Research Foundation
6666 West Quincy Avenue
Denver, CO 80235-3098

Published by:

Awwa Research Foundation
Advancing the Science of Water®

IWA Publishing

DISCLAIMER

This study was funded by the Awwa Research Foundation (AwwaRF). AwwaRF assumes no responsibility for the content of the research study reported in this publication or for the opinions or statements of fact expressed in the report. The mention of trade names for commercial products does not represent or imply the approval or endorsement of AwwaRF. This report is presented solely for informational purposes.

Published by IWA Publishing, Alliance House, 12 Caxton Street, London SW1H 0QS, UK

Telephone: +44 (0) 20 7654 5500; Fax: +44 (0) 20 7654 5555; Email: publications@iwap.co.uk;
Web: **www.iwapublishing.com**

AwwaRF report number 91015F. Originally published by AwwaRF for its subscribers in 2004.
IWA Publishing version published 2005.

ISBN 1 84339 908 3

CONTENTS

TABLES

FIGURES

FOREWORD

The Awwa Research Foundation is a nonprofit corporation that is dedicated to the implementation of a research effort to help utilities respond to regulatory requirements and traditional high-priority concerns of the industry. The research agenda is developed through a process of consultation with subscribers and drinking water professionals. Under the umbrella of a Strategic Research Plan, the Research Advisory Council prioritizes the suggested projects based upon current and future needs, applicability, and past work; the recommendations are forwarded to the Board of Trustees for final selection. The foundation also sponsors research projects through an unsolicited proposal process; the Collaborative Research, Research Applications, and Tailored Collaboration programs; and various joint research efforts with organizations such as the U.S. Environmental Protection Agency, the U.S. Bureau of Reclamation, and the Association of California Water Agencies.

This publication is a result of one of these sponsored studies, and it is hoped that its findings will be applied in communities throughout the world. The following report serves not only as a means of communicating the results of the water industry's centralized research program but also as a tool to enlist the further support of the nonmember utilities and individuals.

Projects are managed closely from their inception to the final report by the foundation's staff and large cadre of volunteers who willingly contribute their time and expertise. The foundation serves a planning and management function and awards contracts to other institutions such as water utilities, universities, and engineering firms. The funding for this research effort comes primarily from the Subscription Program, through which water utilities subscribe to the research program and make an annual payment proportionate to the volume of water they deliver and consultants and manufacturers subscribe based on their annual billings. The program offers a cost-effective and fair method for funding research in the public interest.

A broad spectrum of water supply issues is addressed by the foundation's research agenda: resources, treatment and operations, distribution and storage, water quality and analysis, toxicology, economics, and management. The ultimate purpose of the coordinated effort is to assist water suppliers to provide the highest possible quality of water economically and reliably. The true benefits are realized when the results are implemented at the utility level. The foundation's trustees are pleased to offer this publication as a contribution toward that end.

Walter J. Bishop
Chair, Board of Trustees
Awwa Research Foundation

James F. Manwaring, P.E.
Executive Director
Awwa Research Foundation

ACKNOWLEDGMENTS

The majority of the experimental work completed at the Metropolitan Water District of Southern California (MWD) was performed by Daffodil F. Robles. The authors thank other staff at MWD for a variety of tasks: Connie Chou for operating the UV equipment and measuring UV doses; Ricardo De Leon for technical advice; Patrick Hacker for being available for advice, overseeing the laboratory's standard operating procedures file, coordinating primer synthesis and QA/QC of molecular techniques; Katrin Hanley for assistance with amplicon purification and sequencing; Anne Johnson for continuous development and improvement of the cell culture infectivity assay and for administering the laboratory's purchasing program; and Lynn Kelemen for project and contract administration.

The authors thank Nicolas Ronquillo at Texas Agricultural Experiment Station, El Paso, Texas, for performing quantitative sequence detection assays. We also thank Zia Bukhari (American Water, Belleville, Ill.) for performing an evaluation of an endonuclease site-sensitive assay.

The following water utilities are thanked for their support of the project: Irvine Ranch Water District, Irvine, Calif.; Orange County Water District, Fountain Valley, Calif.; West Basin Municipal Water District, Carson, Calif.; Central Basin Water District, Carson, Calif.; San Diego County Water Authority, San Diego, Calif.

We are also grateful to Cynthia Chappell (University of Texas Health Science Center), Michael Arrowood (Centers for Disease Control and Prevention), Huw Smith (Scottish Parasite Diagnostic Laboratory), Robert Gilman (Johns Hopkins University), and Steve Wright (Moredun Research Institute) for providing *Cryptosporidium* isolates for this study.

Preliminary sequence data was obtained from the University of Minnesota *Cryptosporidium parvum* Genome sequencing project, funded by National Institute of Allergy and Infectious Diseases (NIAID) and accessible at the following website: www.cbc.umn.edu/ResearchProjects/AGAC/Cp/ and the Virginia Commonwealth University/ Tufts University School of Veterinary Medicine *Cryptosporidium* Genome Sequencing Project funded by NIAID and accessible at www.parvum.mic.vcu.edu.

The advice of the Project Advisory Committee—Giovanni Widmer (Tufts University School of Veterinary Medicine, North Grafton, Mass.), Marie-Renee de Roubin (Generale des Eaux, Saint-Maurice, France), and Zia Bukhari (American Water, Belleville, Ill.)—and the help of the Awwa Research Foundation project officers, Albert Ilges and Kenan Ozekin, are also appreciated.

EXECUTIVE SUMMARY

The protozoan parasite *Cryptosporidium parvum* is common in many animal species and oocysts are frequently detected in untreated surface waters. Oocysts are resistant to chlorine disinfection at the concentrations typically applied during drinking water treatment and there are limited chemotherapeutic agents currently available for treatment of infected individuals. Although the disease is usually self-limiting in otherwise healthy patients, it can be fatal for those with weakened immune systems. Consequently, *C. parvum* presents a serious threat to public health and is a significant concern for the water industry. Over 55 documented outbreaks of cryptosporidiosis have been associated with either drinking water or recreational use of water since 1983. The largest waterborne outbreak to date occurred in 1993 in Milwaukee with estimates of the affected population ranging from 15,000 to 400,000 individuals and up to 100 deaths attributed to the outbreak. Various studies have demonstrated that over half of the source water in North America may contain *Cryptosporidium* oocysts.

The primary disinfectants used throughout the water industry in the United States for inactivation of microbial pathogens in drinking water are chlorine and chloramines. However, standard chlorine-based disinfectants have little effect on *C. parvum* oocysts at concentrations typically applied in drinking water treatment plants. In fact, incubation in 10% bleach is a routine pretreatment used to trigger oocyst excystation for cell culture. Consequently, there is an incentive to develop alternative, more effective disinfection strategies such as the use of ozone and ultraviolet (UV) light. UV light is being considered as a disinfectant by the water industry because it appears to be a very effective for disinfecting *Cryptosporidium* oocysts at relatively low doses.

However, it is not known whether *C. parvum* oocysts are killed outright by UV irradiation or merely damaged such that they cannot immediately initiate infection. Many organisms, both prokaryotic (bacteria) and eukaryotic (yeast and other fungi, protozoa, nematodes, plants, amphibians, and mammals) are known to have DNA repair mechanisms such as nucleotide excision repair and photolyase enzymes that act to reduce or eliminate UV-induced damage. Very little is known about the potential for self-repair in *C. parvum* following UV-induced, or any other kind of DNA damage. Oocysts that are sub-lethally damaged by UV light may have the potential to repair and regain infectivity. The ability of *C. parvum* oocysts to reactivate would have serious implications for the use of UV as a disinfectant for the water industry. To overcome oocyst reactivation, it may be necessary to greatly increase applied UV doses or the concentration of post-UV chemical disinfectants, thus increasing the cost of treatment. Also, increasing the concentration of post-UV chemical disinfectants may increase the potential for formation of disinfection by-products, which would partially negate one of the benefits of UV as a primary disinfectant.

RESEARCH OBJECTIVES

The Awwa Research Foundation funded this study to investigate the potential for biological repair in UV-irradiated oocysts. This report presents the results of concurrent lines of investigation to examine whether reactivation occurs following UV disinfection and also to determine if *C. parvum* has the genetic basis for repair. The primary objectives of the study included the following:

1. Determine whether *C. parvum* oocysts of multiple isolates, irradiated with varying doses of UV light, can repair and regain infectivity under light or dark conditions using in-vitro cell culture with a human cell line.
2. Confirm cell culture results with animal infectivity using a neonatal mouse model.
3. Identify possible DNA repair genes in *C. parvum* by performing homology searches with sequence databases, polymerase chain reaction (PCR) assays, and sequencing clones from a genomic library.
4. Measure the extent of UV-induced DNA damage by quantitative PCR.
5. If reactivation is detected, evaluate the effects of both post-UV chemical disinfection on oocyst reactivation and determine the increased UV dosage necessary to inhibit repair.

APPROACH

The project used a combination of empirical experimentation and mining of sequence databases to determine whether *C. parvum* oocysts can repair UV-induced DNA damage and regain infectivity following exposure to UV light. Oocyst infectivity and inactivation were measured using an assay combining in-vitro cell culture and a *C. parvum*-specific reverse transcriptase-polymerase chain reaction (RT-PCR). A limited number of studies used mouse infectivity for corroboration. Infectivity was expressed as the percentage of monolayers that became infected at each oocyst challenge dose and dose response curves were constructed by fitting a least squares regression to a plot of oocyst dose against a logit transformation of the proportional infectivity. The 50% infectious dose (ID_{50}) values were calculated by solving the logistic regression for a logit value of zero. Inactivation was measured by exposing oocysts of five zoonotic isolates (*C. parvum* genotype 2) to varying dosages of UV light and then comparing the infectivity, as reflected by the ID_{50}, of UV-exposed oocysts to untreated control oocysts.

Since DNA repair is an active metabolic process, it is unlikely that the dormant oocyst stage of the parasite will be able to undergo repair. Therefore, this study focused on activated oocysts, released sporozoites, and the later stages of the life cycle. If repair does occur, it is likely to take place in an activated sporozoite before the formation of a trophozoite, since transformation to the trophozoite stage requires active gene expression. The potential for oocyst reactivation was studied by inoculating UV-exposed oocysts onto HCT-8 cell monolayers in the presence and absence of an array of possible reactivation conditions, including incubation in sodium hypochlorite and sodium taurocholate at 37°C for various periods of time following holding periods of up to 24 h under fluorescent white light or in complete darkness at room temperature. Cell monolayers were also incubated for up to five days following inoculation with UV-exposed oocysts instead of the usual 48 h.

DNA repair genes were initially identified in the genome of *C. parvum* by comparison of known repair genes in other organisms with existing nucleic acid sequence databases. Following identification, UV repair genes were isolated from an array of isolates and characterized by PCR, cloning, and sequencing. In addition, the extent of UV-induced DNA damage was assessed by *C. parvum*-specific and arbitrary quantitative sequence detection, and the application of a thymine dimer-specific antibody.

CONCLUSIONS

The overall conclusion from this investigation is that, although *C. parvum* possesses all of the genes necessary for repair of UV-induced DNA damage by nucleotide excision repair, UV disinfected oocysts will not reactivate under typical treated water storage conditions. Consequently, with respect to *C. parvum*, UV is confirmed as an efficient disinfectant for the water industry. Since typical UV dosages in full-scale UV treatment practice are much higher than the relatively low dosages that result in more than 3-$\log_{10}$ (>99.9%) inactivation, it can be concluded that oocysts in drinking water that are exposed to the full UV dosage will not pose a threat to public health. The specific conclusions of the project are:

1. Oocysts of *Cryptosporidium parvum* were readily inactivated at UV doses well below the 40 mJ/cm^2 typically used operationally in full-scale UV water treatment plants.
2. Inactivation levels at 6 mJ/cm^2 of medium pressure UV light were 2.4-$\log_{10}$ (99.6%) for the Iowa isolate, 2.2-$\log_{10}$ (99.4%) for the Moredun isolate, 2.3-$\log_{10}$ (99.5%) for KSU-1, 1.8-$\log_{10}$ (98.2%) for TAMU, and 2.8-$\log_{10}$ (99.8%) for the Maine isolate.
3. Within the limits of data variability, all isolates appeared to have a similar response to UV light and the inactivation data displayed strong agreement with previous studies.
4. There was no difference in *C. parvum* inactivation between low-pressure and medium-pressure UV lamps.
5. The levels of inactivation displayed by oocysts incubated under various repair conditions were generally no lower than oocysts that were inoculated onto cell monolayers immediately following exposure to UV. Therefore, the conclusion of this study is that *C. parvum* oocysts cannot repair UV-induced damage sufficiently to allow them to regain infectivity following exposure to UV light.
6. All of the major genetic components of the nucleotide excision repair complex, the most common repair mechanism for UV damaged cells, were identified in the genomes of *C. parvum* (previously referred to as zoonotic or genotype 2 *C. parvum*) and *C. hominis* (previously referred to as anthroponotic or genotype 1 *C. parvum*).
7. At least three of the UV repair genes were present in multiple *C. parvum* and *C. hominis* isolates and these genes were expressed during infection of cell cultures.
8. Photolyase, cyclobutane pyrimidine dimers-specific DNA glycosylase, and UV-specific endonuclease genes were not identified in *C. parvum* indicating that nucleotide excision repair may be the only UV repair mechanism available, even though it does not appear to function sufficiently to allow oocyst reactivation.
9. The level of UV-induced DNA damage could be assessed by both molecular (quantitative PCR) and antibody-based methods indicating the potential for measuring the effects of UV light without performing infectivity assays.
10. The consistency of the infectivity and inactivation data generated by this project demonstrates the practical application of cell culture-based infectivity assays within the field of microbiological water quality.
11. The successful identification of UV repair genes in *C. parvum* and *C. hominis* highlights the utility of a genomic/bioinformatic approach, making use of the exponentially increasing amount of genomic data available in databases, to address microbiological issues within the water industry.

RECOMMENDATIONS

The water industry should consider moving forward with the implementation of UV-based technologies for disinfecting drinking water. Although there are still issues to be resolved, such as the most appropriate UV dosage, configuration of lamps, type and concentration of secondary disinfectants, potential reactivation of UV disinfected *Cryptosporidium* oocysts should not be part of these considerations. Specific recommendations based on the results of this project include:

1. The water industry should not consider repair of UV-induced damage in *C. parvum* when making decisions on the implementation of full-scale UV treatment.
2. Similar repair studies should be conducted with *C. hominis* (genotype 1 *C. parvum*) to ensure that it too cannot regain infectivity following UV disinfection.
3. In-vitro cell culture should be considered as a practical and proven alternative technology for assessing the infectivity and inactivation of *Cryptosporidium*, although the utility of the method with disinfectants other than UV is yet to be demonstrated. Cell culture should also be used to address issues of infectivity, survival, and disinfection of a broader range of pathogens that may be present in water, e.g., *Giardia duodenalis*, microsporidia, and emerging pathogens.
4. The water industry should partner with environmental and regulatory agencies in furthering studies into the genomic/bioinformatic approach for addressing microbiological questions that can not be answered using the traditional "agar plate" method.

FUTURE RESEARCH

All experiments conducted to date have failed to detect any post-UV parasite development beyond the time of excystation when in-vitro or in-vivo developmental assays are employed, although we have demonstrated the presence of NER genes in both *C. parvum* and *C. hominis*. In addition, it has been reported that *C. parvum* can repair thymine dimers. Therefore, UV-irradiated oocysts may be able to repair UV-induced DNA damage but can not regain infectivity. These findings naturally lead to questions such as: Why does the organism carry the genes if it does not or can not use them? and What is missing from *Cryptosporidium*'s repair machinery that prevents reactivation? Even though oocysts carry NER genes, critical regulatory genes may be missing or contain mutations that prevent efficient operation of the repair mechanism. Because all of the UV repair gene sequences we identified appeared to be complete and none contained premature termination codons, it is likely that these genes are transcribed and translated and that the proteins have specific function(s) within the parasite. We demonstrated that three of the genes (CpRAD1, CpRAD2, and CpRAD3) are expressed during infection but other genes may not be expressed. Although many excision repair enzymes have alternate functions within cells, some such as RAD4 apparently do not. Further studies should be conducted to compare the *C. parvum* RAD genes with those in organisms that can reactivate following UV exposure. Also, more detailed studies of gene expression, gene complementation, and protein function will help increase our understanding of the biological basis of DNA repair in *Cryptosporidium* and the lack of reactivation.

Most of the studies conducted to date have looked for evidence of post-UV reactivation in *C. parvum* by looking for an increase in infectivity following a particular set of repair conditions, relative to UV-irradiated oocysts that were not exposed to these repair conditions. The absence of such an increase in infectivity has been interpreted as an inability to repair UV-induced damage. However, it is possible that oocysts may regain infectivity (since they have the necessary genetic components) but the conditions used to study reactivation are restrictive and some vital component that would permit the parasite to regain infectivity is missing from the assays. Therefore, further research should be conducted on developing infectivity assays that optimize the conditions for potential reactivation. It would be beneficial to include human volunteer studies in this work since all of the UV inactivation and repair studies to date have been conducted using surrogate methods, e.g., cell culture or mouse models. The human intestine may provide the ideal environment for post-UV repair, although it is unlikely.

Although we have identified the same suite of UV repair genes in *C. hominis* (previously genotype 1 or anthroponotic *C. parvum*), the two organisms are recognized as separate species with differences in genetics, physiology, and host-range. Therefore, *C. parvum* inactivation and reactivation data can not be safely extrapolated to *C. hominis*. So, to provide an additional measure of safety and confidence in full-scale applications of UV disinfection within the water industry, more detailed studies should be performed to measure the response of *C. hominis* to UV light, the level of inactivation, and the potential for reactivation.

CHAPTER 1
INTRODUCTION

The protozoan parasite *Cryptosporidium parvum* is common in many animal species and oocysts are frequently detected in untreated surface waters (rivers and lakes). Oocysts are resistant to chlorine disinfection at the concentrations typically applied during drinking water treatment and there are limited effective chemotherapeutic agents currently available for infected individuals. In developing regions of the world, endemic levels of infection and the resulting persistent diarrhea in young children may lead to long-lasting impairment in growth and cognitive function (Guerrant et al., 1999). Although the disease is usually self-limiting in otherwise healthy patients, it can be fatal for those with weakened immune systems. Consequently, *C. parvum* presents a serious threat to public health and is a significant concern for the water industry.

CRYPTOSPORIDIUM IN DRINKING WATER

More than 50 outbreaks of cryptosporidiosis have been associated with drinking water or recreational use of water worldwide (Fayer et al., 2000). The largest waterborne outbreak to date in Milwaukee (1993) affected 15,000 to 400,000 individuals, according to differing estimates (Hunter and Syed, 2001; MacKenzie et al., 1994) and up to 100 deaths were attributed to the outbreak. Over the last 10 years, continued detection of *Cryptosporidium* oocysts in treated drinking water has led to boil water advisories affecting millions of people worldwide. A review of eight studies assessing the occurrence of *Cryptosporidium* spp. in untreated source waters in the United States demonstrated an average proportion of river, lake, and well water samples that were contaminated with oocysts ranging from 9 to 100% (Rose et al., 1997). Water samples that were impacted by domestic and agricultural waste had oocyst concentrations as high as 5,800/L (Madore et al., 1987). A large survey of North America spanning 1988-1993 reported that 60.2% of all samples (n=347) were positive for *Cryptosporidium* oocysts (LeChevallier and Norton, 1995). A similar study in Canada demonstrated lower levels of contamination with oocysts detected in 6.1%, 4.5%, and 3.5% of raw sewage, untreated source water, and treated drinking water, respectively (Wallis et al., 1996).

It is now recognized that *Cryptosporidium parvum* is actually a species complex; genotype 1 or anthroponotic oocysts have been isolated almost exclusively from humans while genotype 2 or zoonotic isolates have been detected in a wide range of animals including humans. Based on genetic, physiological, and host animal differences, and the apparent lack of recombination between the two genotypes, they have recently been classified as separate species. The zoonotic genotype retained the name *C. parvum* while the anthroponotic genotype was named *C. hominis* (Morgan-Ryan et al., 2002). There is considerable strain level variation in the genomic sequences within each species (Figure 1.1). *C. parvum* and *C. hominis* oocysts (or organisms morphologically indistinguishable from *C. parvum* and *C. hominis*) have been reported in at least 152 species of mammals including humans (Fayer et al., 2000). For many years it was thought that *C. parvum* was the only species of *Cryptosporidium* that was infectious for humans. However, other species or genetically distinct isolates of *Cryptosporidium* that have recently been isolated in human infections are *C. canis*, *C. felis*, *C. meleagridis*, and *C. muris* (Fayer et al., 2001; Gatei et al., 2002; Morgan et al., 2000; Morgan-Ryan et al., 2002; Pedraza-Diaz et al., 2001; Pieniazek et al., 1999; Xiao et al., 2001).

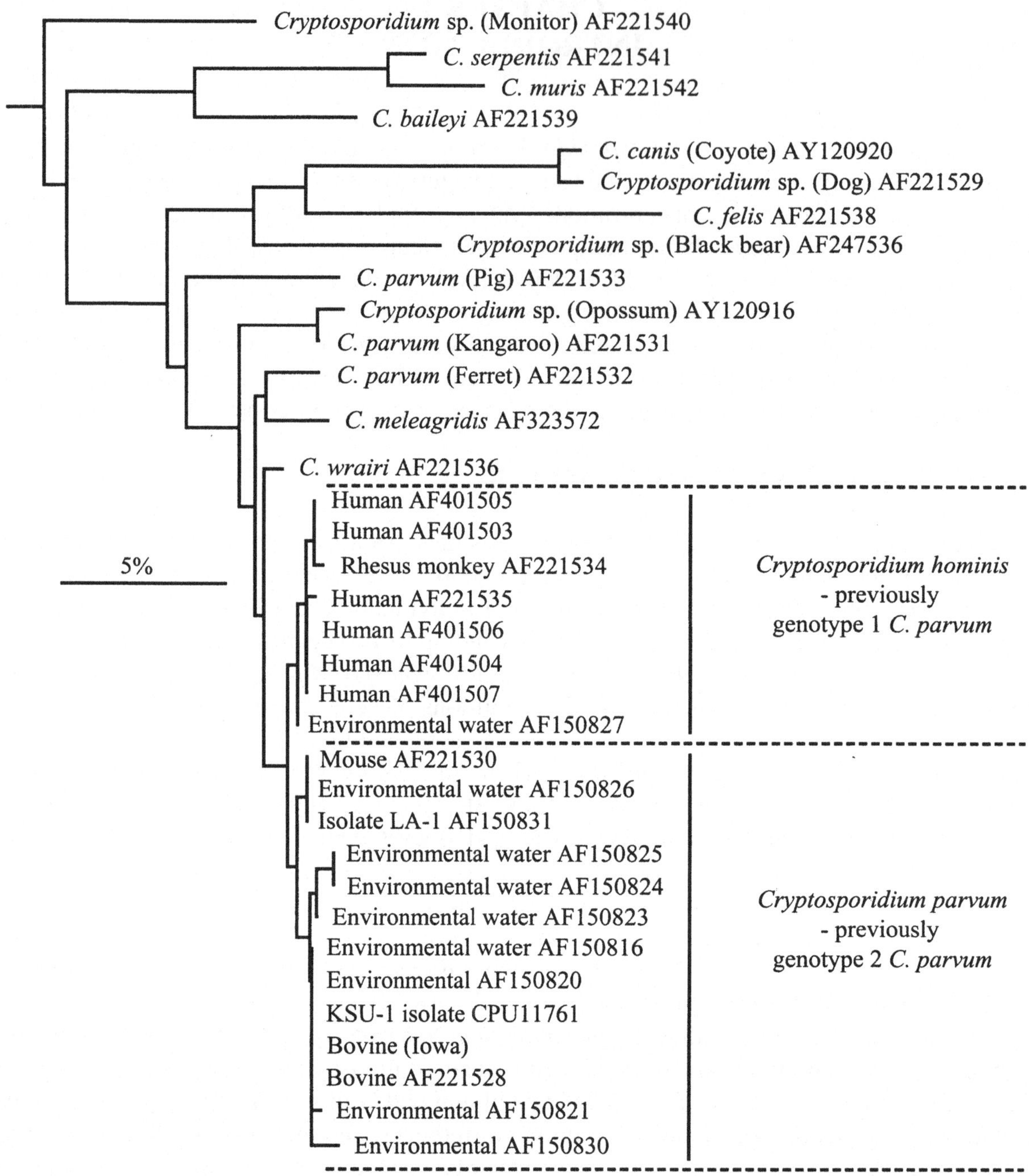

Figure 1.1 Heterogeneity within the 70 kD heat shock protein gene of the *Cryptosporidium* genus. The alignment covered the 361-bp fragment used for detection of infectivity in cell culture in this project (base positions 2423 to 2784, based on sequence CPU11761). Sequences were obtained from GenBank (accession numbers are displayed) or were generated at MWD (Iowa). The corresponding sequence of *Plasmodium falciparum* was used as the outgroup. Distance calculations and tree construction were performed using the Kimura 2-parameter and neighbor joining methods, respectively.

The primary disinfectants used throughout the water industry in the United States for inactivation of microorganisms in drinking water are chlorine and chloramines. However, standard chlorine-based disinfectants have little effect on *C. parvum* oocysts at concentrations typically applied in drinking water treatment plants. In fact, incubation in 10% bleach (0.5% sodium hypochlorite) is a routine pretreatment used to enhance excystation for cell culture (Upton et al., 1994). Some estimates of the chlorine CT (product of contact time and concentration) for *C. parvum* oocysts are as high as 7,200 mg.min/L for >99% inactivation (Korich et al., 1990) compared to a typical value of 0.04 mg.min/L for *Escherichia coli*. Although chlorine dioxide is a more effective disinfectant than free chlorine, CT values of 75 – 1,000 mg.min/L were necessary to achieve 2-$\log_{10}$ (99%) inactivation (Chauret et al., 2001), values that are much higher than concentrations typically used in drinking water treatment. Therefore, alternative disinfectants such as ozone, ultraviolet (UV) light, pulsed white light, ionizing radiation, and bromine have been investigated. Possibly the most promising of these alternatives is UV irradiation. UV has been used for many years to disinfect water intended for pharmaceutical, medical, and food preparation purposes and it is used routinely in some parts of Europe for disinfection of drinking water.

UV INACTIVATION OF *CRYPTOSPORIDIUM PARVUM*

Recent studies have demonstrated that UV light is very effective for inactivation of *C. parvum* oocysts (Table 1.1). Clancy et al. (1998) used a mouse infectivity assay to demonstrate >4-$\log_{10}$ (>99.99%) inactivation using a high volume (775 – 1,550 L/min) flow-through UV irradiation device with a dose of ~8,700 mJ/cm^2. However, further studies demonstrated much higher rates of inactivation. A bench-scale medium-pressure lamp equipped with a collimating tube provided >4-$\log_{10}$ (>99.99%) inactivation at dose of ~80 mJ/cm^2 and a second study with a flow-through device demonstrated 3.9-$\log_{10}$ (99.988%) inactivation in mouse infectivity studies with a dose of 19 mJ/cm^2 (Bukhari et al., 1999). In additional work, UV doses of 6 – 9 mJ/cm^2 provided >3.5-$\log_{10}$ (99.97%) inactivation and 10 mJ/cm^2 of medium pressure UV provided 3.2-$\log_{10}$ (99.95%) inactivation (reviewed by Hargy et al., 2000). A mouse assay demonstrated 2- to 3-$\log_{10}$ (99% – 99.9%) inactivation at 10 to 25 mJ/cm^2, respectively (Craik et al., 2001). Cell culture-based methods combined with fluorescent antibodies demonstrated 3-$\log_{10}$ (99.9%) inactivation at a dose of 3 mJ/cm^2 of low pressure UV (Shin et al., 2001) and >3-$\log_{10}$ (>99.9%) inactivation at 3 mJ/cm^2 (Zimmer et al., 2001). The Metropolitan Water District of Southern California (MWD) operates a large-scale UV treatment research unit (11.6 million liters per day) and has conducted extensive evaluations of the efficacy of UV light for inactivating *C. parvum* oocysts, using bench-scale equipment and an in-vitro cell culture assay to measure infectivity and inactivation (Mofidi et al., 2001; Rochelle et al., 2002). Polychromatic UV doses of 3 mJ/cm^2 and 6 mJ/cm^2 provided an average of 1-$\log_{10}$ (90%) and 2-$\log_{10}$ (99%) inactivation of the Iowa isolate of *C. parvum*, respectively (Mofidi et al., 2001). This latter study also demonstrated that there was no significant difference between medium-pressure and pulsed-UV lamps for inactivation of *C. parvum*. A conservative interpretation of the combined inactivation data, based on the lower 95% prediction limit of the UV dose response curve indicated that 1-$\log_{10}$ (90%) and 2-$\log_{10}$ (99%) reductions in *C. parvum* infectivity were achieved with 7.5 and 11 mJ/cm^2, respectively.

All of the recent studies show good overall agreement in the level of inactivation afforded by UV and demonstrate the sensitivity of *C. parvum* to UV disinfection. In comparison, 3-$\log_{10}$ (99.9%) inactivation of many viruses is achieved with doses ranging from 11 mJ/cm^2 for Hepatitis

Table 1.1
Recent literature on UV inactivation of *Cryptosporidium parvum*

Isolate	Assay method	Example inactivation	UV dose (mJ/cm^2)	Repair conditions	Reference
Iowa	CD-1 mice	>4-log	41	Repair not studied	Bukhari et al., 1999
Iowa	CD-1 mice	3.9-log	8	Repair not studied	Clancy et al., 2000
Iowa	CD-1 mice	3.1-log	11.9	Repair not studied, but UV resistance posited as possible explanation for "tailing"	Craik et al., 2001
Iowa	Cell culture/ RT-PCR	3.3-log	9.4	Repair not studied	Mofidi et al., 2001
HNJ-1	SCID mice	4-log	9.5	Oocysts incubated at 20°C for up to 2 h in light and 24 h in dark	Oguma et al., 2001
Iowa	BALB/c mice Cell culture/IFA	1.6-log 1.7-log	2 2	Oocysts incubated at 25°C and 37°C for up to 4 h in light and dark	Shin et al., 2001
Iowa	Cell culture/IFA	>3.6-log	3	Bleach treated oocysts incubated for up to 72 h in light and dark at 5°C and 25°C	Zimmer et al., 2001

A to 90 mJ/cm^2 for Adenovirus 40. Doses for 3-$\log_{10}$ (99.9%) inactivation of typical enteric bacteria range from 2 mJ/cm^2 for *Shigella dysenteriae* to 9 mJ/cm^2 for *Salmonella enteritidis*.

However, it is not known whether *C. parvum* oocysts are killed outright by UV irradiation or merely damaged such that they cannot immediately initiate infection. Earlier studies of UV exposure using surrogates such as in-vitro excystation and vital dyes as measures of viability indicated that much higher doses of UV (up to 8,700 mJ/cm^2) were necessary to achieve >2-$\log_{10}$ (>99%) inactivation (Campbell et al., 1995). Consequently, if oocysts are sub-lethally damaged by UV, they may have the potential to self-repair and regain infectivity. The ability of *C. parvum* oocysts to reactivate would have serious implications for the use of UV as a disinfectant for the water industry. Should oocysts have a repair capability, it may be necessary to greatly increase applied UV doses or the concentration of post-UV chemical disinfectants, thus increasing the cost of treatment. Many organisms, both prokaryotic (bacteria) and eukaryotic (yeast and other fungi, protozoa, nematodes, plants, amphibians, and mammals) are known to have DNA repair mechanisms that act to reduce or eliminate UV-induced damage (Nickoloff and Hoekstra, 1998). Very little is known about the potential for self-repair in *C. parvum* following UV-induced, or any other kind of DNA damage.

MECHANISMS FOR REPAIRING UV-INDUCED DAMAGE

The heterocyclic ring structures in DNA and RNA absorb light with a maximum absorbance near 260 nm so it can be anticipated that UV light has an effect on DNA. The most important effect of an organism's exposure to UV irradiation is DNA damage. Such damage can be accurately repaired with no long term consequences for the organism, result in mutations which may be silent or deleterious, or cause death of the organism. Reported biological effects of UV irradiation on protozoa include: lowering the lethal temperature and sensitization to heat; altering mobility and shape; delayed division and excystment of cysts; changes in contractile vacuole activity; reduced respiration; and changes in resistance to hydrostatic pressure (Rothschild, 1999). The primary type of UV-induced DNA damage is the generation of photoproducts between adjacent bases. The major photoproducts are cyclobutane pyrimidine dimers (CPDs) between adjacent thymine residues (thymine dimers) and a pyrimidine-pyrimidone product (6-4 lesions; Rothschild, 1999). Both types of DNA damage are highly toxic and mutagenic because they prevent passage of both DNA replication and transcriptional machinery. Thus, a cell with a UV-induced lesion cannot express the genetic information in the region of the genome affected, nor can it accurately copy its DNA and divide. Such a cell may die, or if it survives, become altered and begin to accumulate genetic changes. Additional UV-induced damage includes: hydroxylation of cytosine and uracil, cytosine-thymine dimers, cross-linking between DNA and proteins, interstrand cross-linking of DNA, and chain breakage or denaturation of DNA (Rothschild, 1999). Since many organisms are exposed to potentially high levels of UV irradiation on a daily basis, the ability to repair damaged DNA is essential for the organism's survival and consequently many organisms, both prokaryotic and eukaryotic, are known to have DNA repair mechanisms that act to reduce or eliminate UV-induced damage. These repair mechanisms include: nucleotide excision repair mediated by a complex of 15 – 20 proteins; photoreactivation mediated by light activated photolyase enzymes; UV-specific endonucleases; and base excision repair mediated by CPD-specific glycosylases (Figure 1.2). In addition, some organisms also have the ability to delay cellular processes once DNA damage is detected. These cell cycle delays are mediated by checkpoint genes and they increase a cell's chances of surviving DNA damage by allowing time for the repair machinery to operate prior to DNA replication (Weinert and Lydall, 1998).

Photoreactivation

Photoreactivation has been demonstrated in a wide range of organisms including bacteria, single cell eukaryotes, and complex multi-cellular organisms. The process is light dependent and mediated by a photolyase enzyme that removes pyrimidine dimers by monomerizing them back to the original pyrimidines. The enzyme first binds to the UV-damaged DNA in darkness. Photolyase is then activated by exposure to visible light (400 – 600 nm), which reduces FAD (one of two chromophores contained in photolyase). Once activated and upon exposure to long-wave UV or short-wave visible light (300 – 500 nm), the second chromophore transfers energy to $FADH^-$ and the pyrimidine dimer is split through electron donation (Yasui and Eker, 1998). This is an efficient and rapid process that does not require cleavage of the backbone of DNA.

Based on similarities in deduced amino acid sequences, all known photolyases fall into two classes: Class I enzymes contain two chromophores and have been detected in bacteria, lower eukaryotes, insects, and mammals; Most class II enzymes contain one chromophore and are found in some bacteria, archae, fish, and marsupials. Although there is considerable variation in the sequences of the proteins, some regions are conserved; in general, there is 25-80% amino acid conservation among different photolyases (Todo, 1999). However, some organisms do not appear to contain a photolyase gene (e.g. *Schizosaccharomyces pombe*, *Haemophilus influenzae*, *Mycoplasma genitalium*) and there have been no reports of photolyase enzymes in coccidial parasites. Since coccidia are obligate intracellular parasites, they may have lost the photolyase gene during their evolution because light of the necessary wavelength for photolyase activity would not have been available.

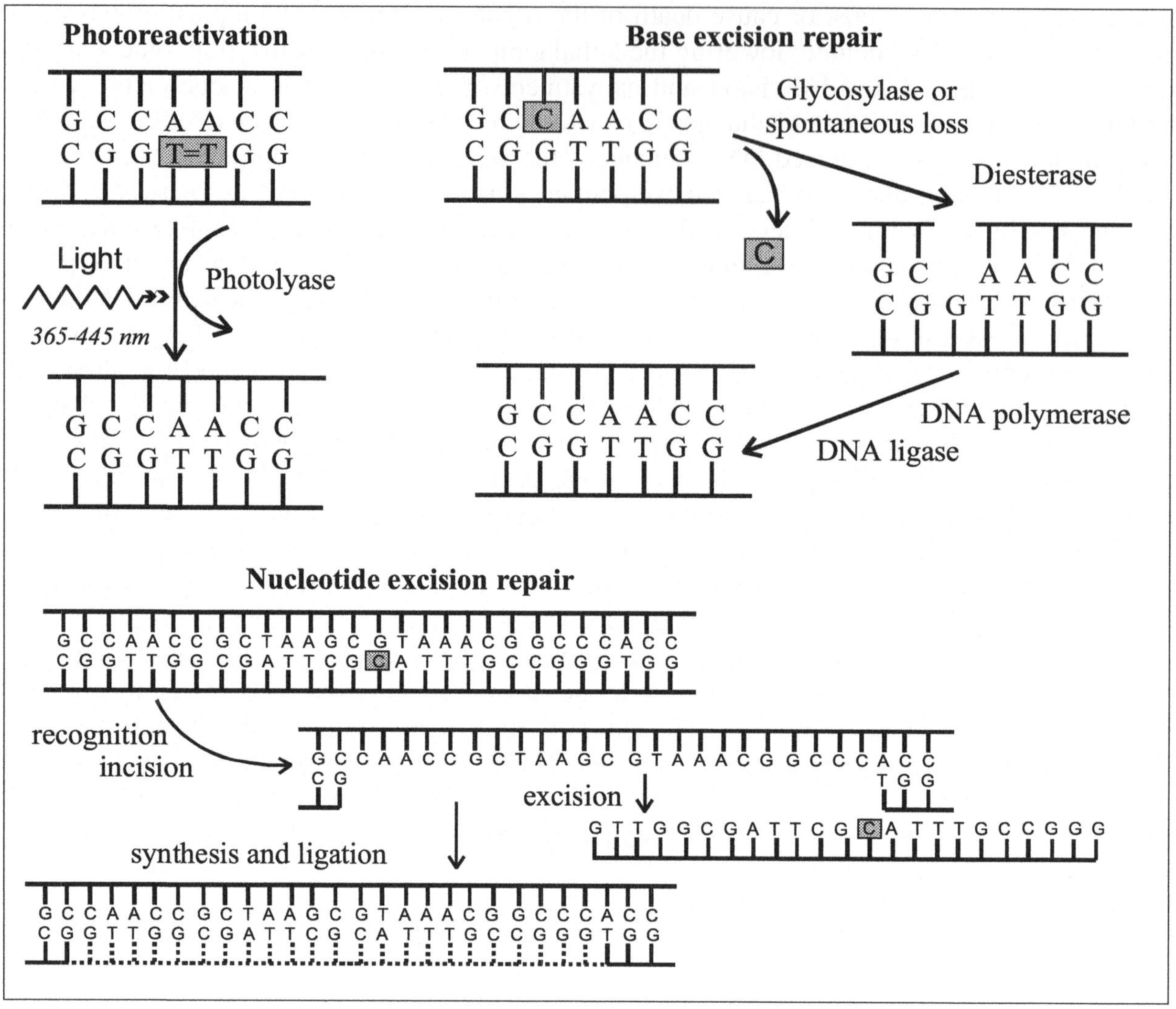

Figure 1.2 Summary of three types of DNA damage repair mechanisms. Boxes indicate damaged nucleotides.

Base Excision Repair

Base excision repair (BER) involves the removal of a damaged base by either a DNA glycosylase or spontaneous loss. Following removal of the baseless sugar by an apurinic/apyrimidinic endonuclease, the repair is completed by DNA polymerase adding a new base and DNA ligase sealing the single-stranded nick. There are many glycosylase enzymes that can initiate BER (e.g. *E. coli* contains up to 8 separate glycosylases with varying substrate specificities). However, BER of UV-induced damage is mediated by only one kind of glycosylase- pyrimidine dimer (PD)-DNA glycosylase. PD-DNA glycosylase recognizes CPDs and makes a cut between the altered base and the deoxyribose sugar to which it is attached. This reaction is specific for CPDs and does not work on other UV-induced lesions. As with other glycosylases, the glycolytic cleavage is followed by removal of the baseless sugar by an AP endonuclease, followed by filling in of the missing nucleotide by a DNA polymerase. PD glycosylases are not widely distributed in nature; they have been

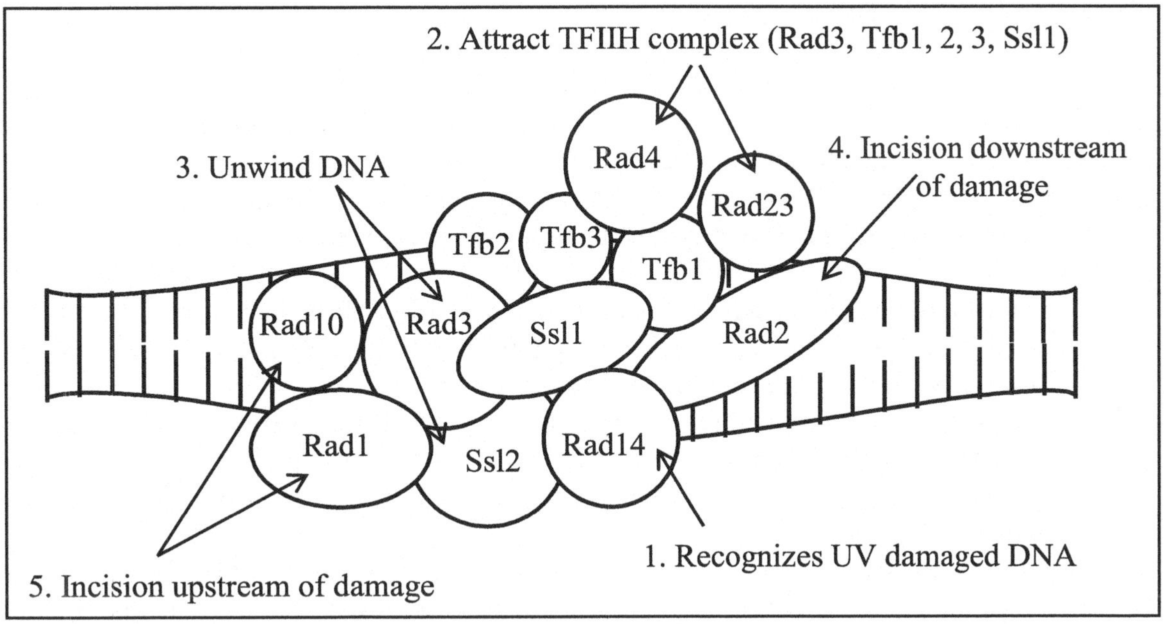

Figure 1.3 Proteins involved in nucleotide excision repair in *Saccharomyces cerevisiae* (adapted from Siede, 1998)

identified only in bacteriophage T4, the bacterium *Micrococcus luteus*, and a virus (PCBV-1) from the green alga *Chlorella* (McCullough et al., 1998). There have been no reports of PD-DNA glycosylases in coccidial parasites.

UV-Specific Endonuclease

Direct-acting endonucleases that cleave 5' to CPDs and 6-4 lesions have been isolated from the fungi *Neurospora crassa* and *S. pombe*, and homologs have been reported in the bacteria *Bacillus subtilis* and *Deinococcus radiodurans* (Yasui and McCready, 1998). These enzymes are not widely distributed in nature and no reports exist to date of UV endonucleases in coccidial parasites.

Nucleotide Excision Repair

Nucleotide excision repair (NER) is a versatile and ubiquitous repair process in which alterations of the DNA double helix are recognized by a multi-protein complex which then makes dual incisions in the strand of the DNA containing the damaged nucleotide(s), resulting in removal of a 24-32 nucleotide fragment. The minimum set of proteins or protein complexes required for damage recognition and the dual cleavage reaction is approximately fifteen in eukaryotes (Figure 1.3). Cleavage is followed by displacement of the damaged fragment and filling in of the gapped region by DNA polymerase. Members of the NER complex include both those proteins uniquely involved in NER and those that have additional functions including acting in transcription.

PROJECT OBJECTIVES

The overall objective of this project was to investigate the potential for repair in UV-irradiated oocysts. Concurrent lines of investigation measured UV inactivation of multiple isolates of *C. parvum*, assessed the ability for post-UV repair in oocysts, assessed whether *C. parvum* has the genetic basis for repair of UV-induced DNA damage, and examined the extent of UV-induced damage in oocysts. The original specific objectives for the project were:

1. Determine whether *C. parvum* oocysts of three different isolates, irradiated with varying doses of UV, can self-repair and regain infectivity under light or dark conditions using in-vitro cell culture with a human cell line.
2. Confirm cell culture results with animal infectivity using a neonatal mouse model.
3. Identify possible DNA repair genes in *C. parvum* by PCR, probe hybridization, sequencing clones from a genomic library, and performing homology searches with sequence databases.
4. Measure the extent of UV-induced DNA damage by quantitative PCR.
5. If self-repair is detected, evaluate the effects of post-UV chemical disinfection on oocyst reactivation and determine the increased UV dosage necessary to inhibit repair.

SIGNIFICANCE AND BENEFITS TO THE WATER INDUSTRY

UV irradiation appears to be an effective disinfectant that may greatly reduce the risks to public health posed by waterborne *C. parvum* and other organisms. In addition to its ability to effectively inactivate many pathogens, UV irradiation also has the advantage of being a non-contact disinfectant: it does not require the physical addition of compounds to the water. The technology has the potential to revolutionize water treatment and full-scale units have been installed in various treatment plants around the world. Many more utilities are likely to install UV treatment because it appears to reduce the likelihood of human infections by waterborne *Cryptosporidium*, and other disinfectant-resistant organisms, without generating the harmful by-products that are a major disadvantage of chemical-based disinfectants. Studies of low-pressure UV demonstrated that no significant changes could be detected in the assimilable organic carbon, biodegradable dissolved organic carbon, or other biodegradable organic matter constituents after dosages of UV light up to 130 mJ/cm^2 (Shaw et al., 2000). Also, the use of UV as a primary disinfectant does not present the potential safety hazard that is associated with use of chlorine gas and other strong oxidants. However, UV is somewhat unique among disinfection agents in that many organisms have the ability to self-repair following exposure and apparent inactivation. This study assessed the ability of *C. parvum* oocysts to repair and regain infectivity following exposure to UV. The information provided by the project may be used by the water industry in making decisions about the design, installation, and operation of UV disinfection systems in drinking water treatment plants.

CHAPTER 2
MATERIALS AND METHODS

All animals used in this research for propagation of oocysts and infectivity assays were housed, handled, and monitored in an accredited animal care facility at either the University of Arizona or Kansas State University in accordance with the Animal Welfare Act of 1985, the Public Health Service Policy of 1986, the Good Laboratory Practice Act of 1978, and the National Academy of Sciences *Guide for the Care and Use of Laboratory Animals* (1996).

OOCYST PRODUCTION AND PURIFICATION

The *Cryptosporidium* isolates used in this investigation are described in Table 2.1. Five of them were categorized as zoonotic or genotype 2 *C. parvum* isolates and the sixth (PERU-RG2) was genotype 1 or *C. hominis*. For infection and UV inactivation experiments, *C. parvum* oocysts were propagated by artificial infection of Holstein calves housed in the University of Arizona Central Animal Facility (accredited by the American Association of Accreditation of Laboratory Animal Care). Calves were fed 2×10^8 oocysts suspended in sterile water within 4 to 12 h of birth, followed 1 – 2 h later by 1.9 L of colostrum. The calves received prophylactic doses of oral Rota/Corona Virus and *Escherichia coli* vaccines. Subsequently, the calves were maintained on a diet providing 2200 kcal/day supplied by a combination of milk replacer, non-fat milk powder, and electrolyte mix. All procedures for calf care and infection were approved by the University of Arizona's Institutional Animal Care and Use Committee. The total daily output from infected calves was collected, screened through graded sieves, and concentrated by centrifugation. Oocysts were purified from fecal material by sequential density gradient centrifugation through sucrose and cesium chloride gradients as described previously (Arrowood and Donaldson, 1996). Purified oocysts were suspended in phosphate-buffered saline (PBS) containing 0.01% Tween 20 and the following antibiotics: penicillin (100 U/mL), streptomycin (100 μg/mL), and gentamicin (100 μg/mL), and stored at 4°C until they were used for infection.

For gene identification studies at Kansas State University, oocysts of the KSU-1 isolate were generated by artificial infection of one day-old male Holstein calves received through the Animal Resource Facility at the KSU College of Veterinary Medicine. Calves were housed individually in stainless steel pens equipped with vinyl-covered gratings overlaying a stainless steel collection pan, in a temperature controlled room on a 12 h light/dark cycle. The collecting pans were cleaned daily. All animals were bottle fed with milk replacer 4 times per day. At 5 days of age fecal samples were examined microscopically to ensure no natural *C. parvum* infections were present. Infection-negative calves were fed 50 mL of PBS containing 2×10^7 viable KSU-1 oocysts. Starting 4 days after infection and continuing through 10 days post-infection, all feces were retrieved from the collecting pans. During the collection period, the calves' diets were supplemented with electrolyte replacement therapy 4 times per day. Fecal samples were passed through a graded series of wire mesh sieves to a final exclusion limit of 80 μm, oocysts were pelleted by centrifugation, and purified by cesium chloride gradient centrifugation. Purified oocysts were washed several times in sterile PBS and oocysts pelleted by centrifugation. For transport between the various laboratories involved in this investigation, oocysts were packed on ice, packaged and labeled as required by the International Air Transport Association Dangerous Goods

Table 2.1
***Cryptosporidium parvum* isolates used in this study**

Isolate	Original host	Source	Reference
Iowa	Cow	H. Moon, National Animal Disease Center, Ames, Iowa	Korich et al., 1990
KSU-1	Cow	S.J. Upton, Kansas State University, Manhattan, Kansas	Khramtsov et al., 1995
Maine	Human	M.J. Arrowood, Centers for Disease Control, Atlanta, Georgia	Millard et al., 1994
Glasgow	Human	H.V. Smith, Scottish Parasite Diagnostic Laboratory, Glasgow, Scotland	H.V. Smith, pers. comm.
TAMU	Horse	C.L. Chappell, University of Texas Health Science Center, Houston, Texas	Okhuysen et al., 1999
Moredun	Deer	S.E. Wright, Moredun Research Institute, Peniculk, Scotland	Blewett et al., 1993
PeruRG2	Human	R. Gilman, Johns Hopkins University, Baltimore, Maryland	R. Gilman, pers. comm.

Regulations and shipped via Federal Express. Chain of custody forms accompanied all shipments of oocysts.

Standard quality assurance and control procedures for most lots of oocysts received at MWD involved microscopic observation to assess the amount of general debris and contaminating microorganisms, and inoculation of nutrient agar and m-endo agar plates (incubated at 37°C for 24 h), Sabaroud-dextrose agar plates (incubated at 28°C for 7 d), and RPMI cell culture media (including antibiotics) to assess the level of bacterial and fungal contamination. Experience has demonstrated that *C. parvum* oocyst preparations are sometimes contaminated with *Escherichia coli* (Table 2.2). Some of these contaminating strains of *E. coli* are resistant to the antibiotics used in the cell culture medium. Any putative positive coliforms on the m-endo plate were identified using API20E strips following the manufacturer's recommendations (Biomerieux, Hazelwood, MO). The presence of antibiotic resistant bacteria in oocyst preparations highlights the importance of a routine QC procedure for oocysts, particularly when they are intended for use in cell culture-based assays.

In-Vitro Excystation of *C. parvum* Oocysts

The in-vitro excystation efficiency was determined for all batches of oocysts used in this investigation (Table 2.2). Oocysts were incubated in acidified Hank's balanced salt solution for 1 h at 37°C followed by 1 h at 37°C in a solution containing 0.8% trypsin and 0.75% sodium taurocholate, and finally 30 min at room temperature in the latter solution. The rate of excystation was determined by microscopic examination to enumerate the number of excysted and unexcysted oocysts.

Table 2.2
Quality control of oocysts used for inactivation and repair experiments

					Contamination		
Isolate	Lot number	Received at MWD	Excystation rate (%)	Age (days)*	m-endo	*E. coli*†	RPMI + Abs‡
Iowa	10226-06	03/13/01	96	20-27	nd	nd	nd
Iowa	10327-01	04/11/01	98	15-29	nd	nd	nd
Iowa	10423-28	05/02/01	96	16-23	–	–	–
Iowa	10514-21	06/01/01	100	14-28	nd	nd	nd
Iowa	10604-11	06/18/01	nd	35	+	–	–
Iowa	10709-14	07/19/01	nd	9-16	–	–	–
Iowa	10806-13	08/22/01	99	21-35	–	–	–
Iowa	10910-15	09/20/01	95	23	–	–	–
Iowa	11203-11	01/03/02	nd	62-230	–	–	–
Iowa	20102-10	01/15/02	98	36	+/+§	+/+§	+/–§
Iowa	20225-06	03/14/02	95	45	+	+	–
Iowa	20422-27	05/02/02	97	11	–	–	–
Iowa	20506-14	05/23/02	96	14-42	–	–	–
Iowa	20722-29	08/13/02	97	28	+	+	–
Iowa	20708-17	07/30/02	97	60	–	–	–
Iowa	20722-29	09/12/02	98	49-70	–	–	–
Iowa	20924-01	10/09/02	96	13	+	–	–
Iowa	21105-13	12/04/02	92	19	–	–	–
Iowa	30210-15	03/06/03	96	32			
Maine	10723-30	08/02/01	nd	7-28	+	+	+
Moredun	10102M-12	01/24/01	98	25	nd	nd	nd
KSU-1	110201HW	11/20/01	95	24-44	–	–	–
TAMU	10212T-19	02/27/01	98	15-29	nd	nd	nd

*Number of days post-shedding at the time of infection.
†Identification of *E. coli* as primary contaminant in oocyst preparation.
‡Growth in RPMI cell culture media containing antibiotics.
§Oocysts were screened for contamination immediately upon receipt and following a second cesium chloride gradient purification. Results are reported as initial screen/second screen.
nd, Not determined.

Enumeration of Oocyst Challenge Doses

Accurate enumeration of oocyst challenge doses is a critical component of measuring oocyst infectivity and inactivation. This ensures that differences in the amount of infectivity or inactivation are real differences and not apparent differences resulting from inaccurate or imprecise enumeration of oocyst doses. Oocyst stock preparations were enumerated using a modified Neubauer hemacytometer to generate a mean of five replicate counts. The challenge doses used for infections were prepared by dilution in PBS and were enumerated by epifluorescence microscopy of oocysts stained with a fluorescein isothiocyanate labeled monoclonal antibody. At the time of UV exposure or cell culture infection, at least four 100 µL aliquots of diluted oocysts were placed on two-well Superstick slides (Waterborne, Inc., New Orleans, La.) or three-well slides (Meridian Diagnostics, Cincinnati, OH) and allowed to dry overnight at room temperature. Immunofluorescence counts were performed on 100 µL aliquots of each challenge dose because this was the same volume that was used to inoculate cell monolayers. Diluted Aqua-Glo GC Direct antibody (Waterborne, Inc.) was applied and incubated at 37°C for 30 min. Following rinsing in PBS and deionized water, slides were mounted (Merifluor mounting medium; Meridian Diagnostics) and examined using a Nikon Optiphot fluorescence microscope with excitation and emission wavelength filters of 450 – 490 nm and 515 nm, respectively. Each challenge dose was calculated as the mean of at least four individual counts. It is generally recognized that averaging 10 individual immunofluorescence counts is an appropriate enumeration method but the amount of microscopy involved in counting 10 wells for each challenge dose for every experiment was prohibitive within this project. Nevertheless, enumeration of oocyst challenge doses during this project using four-well and five-well counts resulted in a generally acceptable level of precision (Figures 2.1 and 2.2). Oocyst challenge doses used throughout this project to infect HCT-8 cell cultures ranged from 6 to 5,050 and the mean coefficient of variation (CV) for all of the challenge doses was 8.5% with a range of 1.1 – 60.4%. Figure 2.2 demonstrates that the actual oocyst dose inoculated onto cell monolayers correlated strongly with the intended dose based on hemacytometer enumeration. The data in Figures 2.1 and 2.2 were derived from 1,066 individual well enumerations (234 four-well counts and 26 five-well counts). Allowing approximately 20 minutes for enumeration of oocysts in each well resulted in 355 hours of microscopic observation. As would be expected, there was an inverse relationship between the enumerated oocyst doses and CV although the relationship was not strong for oocyst doses greater than 500. For these oocyst doses, 25 µL aliquots or 100 µL aliquots of diluted samples were enumerated by immunofluorscence microscopy and the challenge dose was calculated by appropriate multiplication. Therefore, the decreased precision of the higher doses is explained by the error resulting from enumerating sub-samples of oocyst suspensions.

UV IRRADIATION OF OOCYSTS

Oocysts were exposed to UV light using a collimated beam apparatus (Calgon Carbon Corporation, Pittsburgh, Penn.) equipped with a 6.4 cm inside diameter collimating tube below either a medium pressure (1 kW) or a low pressure UV lamp. The low pressure lamp emitted monochromatic radiation with peak output at 254 nm while the medium pressure lamp was polychromatic with emission peaks at 248, 266, 280, 289, and 295 nm (Figure 2.3). Oocysts were suspended at a concentration of $1 \times 10^5 - 5 \times 10^6$/mL in 10 mL of PBS or treated effluent from MWD's Weymouth Filtration Plant in a 60-mm poly-methylpentene petri dish and exposed to a

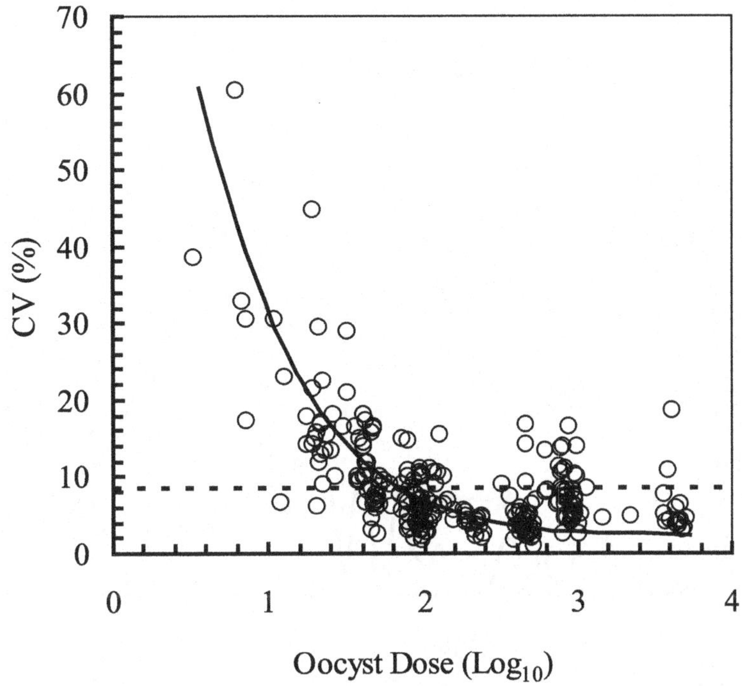

Figure 2.1 Relationship between oocyst challenge dose and the coefficient of variation (CV) for enumeration of oocysts by immunofluorescence. N = 1,066 individual well counts. The horizontal dashed line represents the mean CV (8.5%).

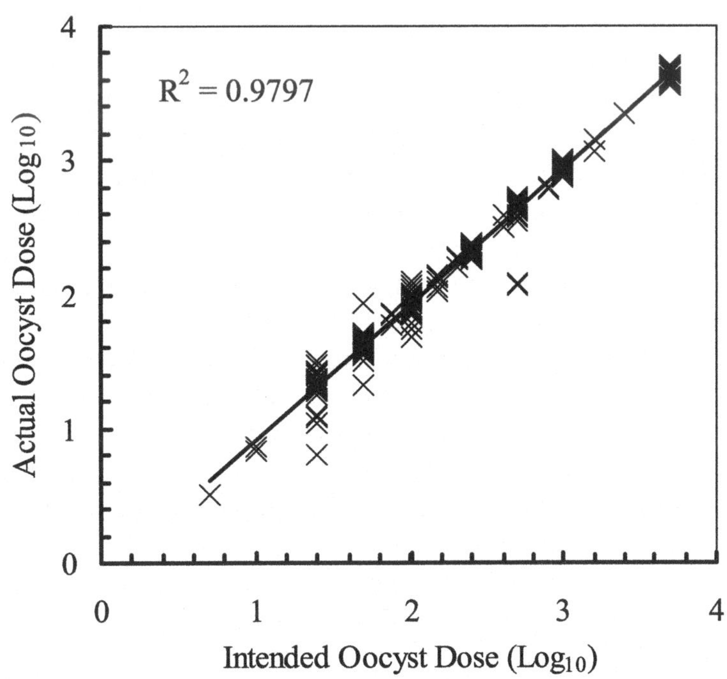

Figure 2.2 Correlation between the intended oocyst challenge dose (based on hemacytometer enumeration) and the actual oocyst dose, based on immunofluorescence enumeration. N = 1,066.

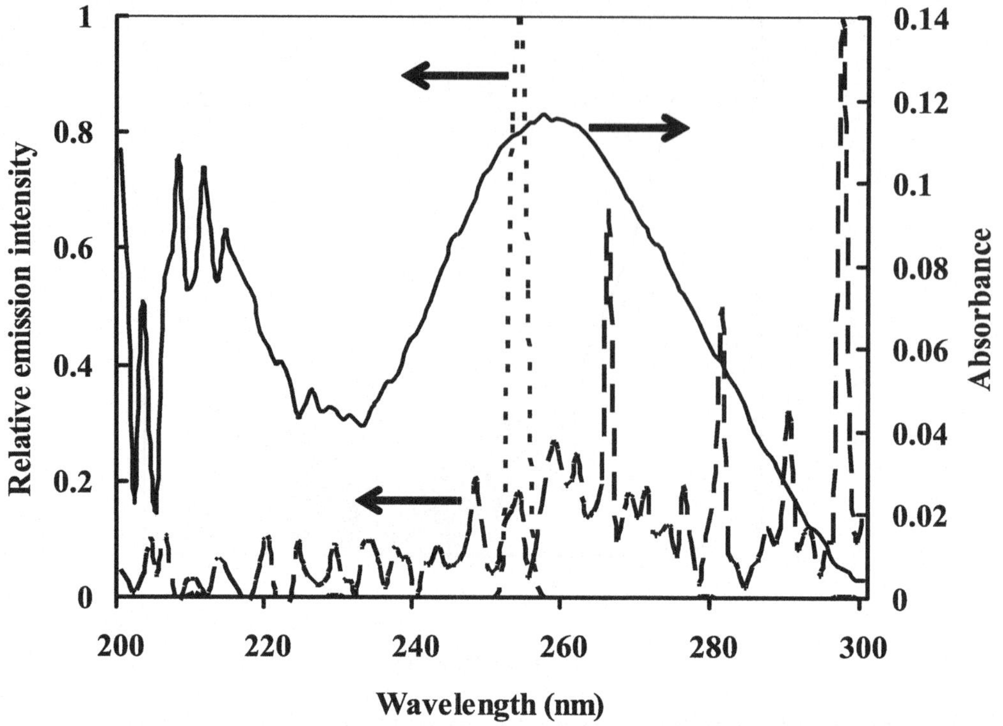

Figure 2.3 Emission spectra of low pressure (- - -) and medium pressure (— —) UV lamps (Linden and Mofidi, 2003) overlaid on the absorbance spectrum of a purified plasmid DNA preparation (——). The arrows indicate the appropriate axis for each spectrum.

variety of UV dosages to achieve a range of inactivation levels. Typical annual averages of physio-chemical characteristics for MWD's filter effluent at this treatment plant are 0.07 NTU (range 0.05 – 0.08), pH 8.13 (range 8.07 – 8.21), and 18.5°C (range 12.3 – 25.3). The samples were continuously mixed using a micro stir bar and magnetic stirrer during UV exposure. Oocysts were exposed for variable lengths of time, at a constant distance, from a constant intensity UV light to achieve the different UV dosages (mJ/cm^2).

The irradiance is a function of the lamp power and the lamp to sample distance. Incident UV irradiance was measured using a radiometer (IL 1700 radiometer, International Light, Newburyport, Mass.) with mathematical correction and confirmed utilizing chemical actinometry to calibrate the dosages delivered to the oocyst suspensions. The incident UV irradiance was measured at the center of the liquid surface of the sample. By making appropriate measurements at positions away from the center, the average incident UV irradiance across the top of the petri dish was obtained and the average UV dose in the 10 mL sample was calculated using the integration of the Beer-Lambert Law through the depth of the liquid sample (Morowitz, 1950) and adapted for polychromatic radiation (Linden and Mofidi, 1999). A more detailed description of the UV apparatus and the methods used to measure UV dosages has been published previously (Mofidi et al., 2001).

Following UV exposure, the oocyst samples were either inoculated immediately into HCT-8 cells or mice to determine the level of inactivation and as zero reactivation controls or exposed to various potential reactivation conditions prior to inoculation.

For some experiments designed to measure the extent of UV-induced DNA damage, oocysts were exposed to medium-pressure UV light using a system similar to the one described above and to low-pressure UV using a cross-linking instrument (Stratagene).

CELL CULTURE PROCEDURES FOR INFECTIVITY STUDIES

The infectivity of non-exposed and UV-exposed oocysts was measured using in-vitro cell culture. The majority of reagents used for cell culture were obtained from Sigma Chemical Co. (St. Louis, Mo.). Human ileocecal adenocarcinoma HCT-8 cells (CCL 244, American Type Culture Collection, Rockville, Md.) were initially grown and maintained in 75 cm^2 and 150 cm^2 flasks. For measuring infectivity and inactivation, cells were grown in 24-well culture plates (Corning Inc., Corning, N.Y.) in RPMI-1640 medium supplemented with 4 mM L-glutamine, 30 mM HEPES, pH 7.3, 100 U/mL penicillin, 100 µg/mL each of streptomycin and kanamycin, and 0.25 µg/mL amphotericin B. Cell culture growth medium was supplemented with 10% fetal bovine serum (FBS; HyClone, Logan, UT) for maintenance of uninfected cells and 2% FBS for cells following inoculation with oocysts. Cells were incubated in a humidified incubator at 36°C in an aerobic atmosphere containing 5% CO_2. HCT-8 cells were inoculated into wells to achieve an initial density of 1×10^5 cells/cm^2, and cultures incubated until the monolayers reached 100% confluence. Untreated or UV exposed oocysts were resuspended in cell culture maintenance media and inoculated in 100 µL aliquots onto 100% confluent monolayers. In some instances, oocysts were incubated in 0.75% sodium taurocholate (prepared in RPMI medium) at 37°C for 10 min to trigger excystation and washed with PBS prior to inoculation of monolayers. Following incubation of inoculated cell monolayers for 48 h at 36°C in 5% CO_2, *C. parvum* infection was detected by a reverse transcriptase polymerase chain reaction (RT-PCR) assay of extracted RNA.

Detection of Infection by RT-PCR

Following incubation of inoculated monolayers for 48 h, infection in HCT-8 cells was detected by RT-PCR amplification of a *C. parvum*-specific 361 bp fragment of mRNA transcribed from the heat shock protein 70 (*hsp*70) gene using a modification of a previously described procedure (Rochelle et al., 1997). This method does not attempt to use RT-PCR detection in a quantitative manner, but simply as a measure of presence or absence of infection in each of the multiple cell monolayers that were infected with each challenge dose of oocysts. The approach is analogous to the standard procedure used to quantify infectivity in mice in which infectivity is expressed as the proportion of animals that become infected.

All glassware used for RNA extraction reagents was baked at 180°C overnight, and all reagents were treated for at least 8 h at room temperature with 1% diethyl pyrocarbonate (DEPC) or made with DEPC-treated sterile distilled water to destroy contaminating RNA degrading enzymes. DEPC was removed from water and reagents by autoclaving at 121°C for 30 min.

Total RNA was extracted using either S.N.A.P. Total RNA isolation (Invitrogen, Carlsbad, CA) or RNeasy Kits (Qiagen, Valencia, CA) following the manufacturer's recommendations. Digestion of extracted RNA with DNase (Ambion, Austin, TX) was included in the procedures to ensure that the final RT-PCR signal resulted from amplification of mRNA derived from sites of active infection and not from carryover of DNA that may be co-extracted from inactive oocysts remaining on the cell monolayers. The final RNA pellet was resuspended in 100 µL of DEPC-treated sterile deionized water. Logistical reasons sometimes prevented the immediate extraction of RNA from cell monolayers following incubation so cell monolayers were preserved in RNA Later (Ambion) and stored at -20°C. In-house comparisons demonstrated no difference in RT-PCR signals and infectivity levels resulting from immediate RNA extraction or extraction from RNA Later-preserved cells. The reverse transcription (RT) and polymerase chain reactions (PCR) were performed separately in all cases. RT reactions were performed using an oligo $(dT)_{16}$ primer, MuLV reverse transcriptase, 1 U of RNase inhibitor per reaction (all supplied by Applied Biosystems, Foster City, CA), 200 μM each of dATP, dCTP, dGTP, and dTTP (Roche, Indianapolis, IN), and 10 μL of extracted RNA in PCR buffer II (Applied Biosystems) containing 5 mM $MgCl_2$ in a 20-μL reaction volume. Incubation conditions for the RT reactions were 10 min at 25°C, 60 min at 42°C, and 5 min at 99°C. Amplification reactions were mediated by 0.025 U/μL of Platinum Taq DNA polymerase (Invitrogen) and 0.25 μM of the *C. parvum*-specific primers cphsp2423F and cphsp2764R (Table 2.3) with an annealing temperature of 53°C. Amplification reactions contained 10 mM Tris-HCl (pH 8.3), 50 mM KCl, 1.5 mM $MgCl_2$, 200 μM each of dATP, dCTP, dGTP, and dUTP, 0.01 U/μL of uracil DNA glycosylase (UDG; Roche), and 10 μL of the cDNA produced by the RT reaction. UDG and dUTP instead of dTTP were included as a quality control measure to prevent carryover contamination of PCR products from earlier reactions. Reactions were performed in a final volume of 50 μL in 96-well plates (Phenix Research Products, Hayward, CA) using ABI 9700 (Applied Biosystems) and PTC-200 (MJ Research) thermal cyclers. Incubation conditions for the amplification reactions were 10 min at 24°C to allow the UDG to digest any carryover contamination from previous reactions, 10 min at 95°C to simultaneously inactivate the UDG and activate the Platinum Taq polymerase, 40 cycles of 45 sec at 94°C, 45 sec at 53°C, and 90 sec at 72°C followed by a single cycle of 5 min at 72°C and finally cooling to 4°C.

Amplicons were visualized by agarose gel electrophoresis and ethidium bromide staining using standard conditions (Sambrook et al., 2001). For selected experiments amplicon identity was confirmed following gel electrophoresis by hybridization with a digoxigenin (DIG)-labeled oligonucleotide probe specific to the *C. parvum* hsp70 amplicon (cphsp2475, Rochelle et al., 1997). DIG-labeled probe-amplicon hybrids were detected using an anti-DIG antibody conjugated to alkaline phosphatase and a chemiluminescent substrate (CDP Star; Roche).

Quality Control for Cell Culture and RT-PCR Procedures

All handling, maintenance and culturing of stock and uninfected cells was performed in a dedicated cell culture facility equipped with a biological safety cabinet, laminar flow hood, and cell culture dedicated incubators. Uninfected and *C. parvum*-infected cell monolayers were maintained in separate incubators in different rooms to minimize the risk of false positive infections. Routine quality control (QC) procedures for the cell culturing facility included assessing the sterility of all media and supplements used for culturing using Sabaroud-Dextrose agar, blood agar, brain heart infusion broth, and thioglycollate broth (Hardy Diagnostics, Santa Maria, CA). Approximately 5% of all the prepared cell culture media was used for QC purposes. If any batch of prepared cell culture media tested positive for bacterial or fungal growth, it was discarded along with any cells that had been grown in it.

Each experiment to measure oocyst infectivity, UV inactivation, and repair included at least two uninfected cell culture wells as negative infection controls, monolayers inoculated with 1,000 heat killed oocysts (70°C for 30 min) as infection and carryover controls, and monolayers inoculated with approximately 1,000 untreated oocysts as positive controls. Quality control also included tracking all reagents used for the RT-PCR detection assays through assigned lot numbers.

Table 2.3
PCR primers and probes used in this investigation

Primer	Sequence, 5'-3'	Target gene	Amplicon size (bp)	Annealing temp.	Reference
cphsp2423F	AAATGGTGAGCAATCCTCTG	hsp70*	361	53°C	Rochelle et al., 1997
cphsp2764R	CTTGCTGCTCTTACCAGTAC				
cphsp2475†	CCATTATCACTCGGTTTAGA				
CpRAD1F	TAGCGTCTGGAGTGTCG	CpRAD1	369	60°C	This study
CpRAD1R	GCGGTGAATGTAATGAGAGT				
CpRAD2F	ACCCGACCTAGAACCA	CpRAD2	304	60°C	This study
CpRAD2R	AGCTTCTGCTTCACCT				
CpRAD3F	GCACTCGTACTGTGGC	CpRAD3	345	60°C	This study
CpRAD3R	TTGCATCCCCCTCACT				
CPHSPT2F	TCCTCTGCCGTACAGGATCTCTTA	hsp70*	345	60°C	Di Giovanni et al., 2002
CPHSPT2R	TGCTGCTCTTACCAGTACTCTTATCA				
CPHSP2P2†	FAM‡-TGTTGCTCCATTATCACTCGGTTTAGA-TAMRA§				
GPD62	CAATGCCCGA	random	NA**	40°C	Deng and Cliver, 1998
GPD63	GTCCGGAGAA	random	NA**	40°C	Deng and Cliver, 1998

*Heat shock protein 70 kDa gene.
†Oligonucleotide probe sequences.
‡6-carboxyfluorescein.
§6-carboxy-tetramethylrhodamine.
**Not applicable since multiple arbitrary fragments were amplified.

Negative controls for the RT-PCR assay included omitting the reverse transcriptase from the RT reaction to ensure that amplified product originated from mRNA and not DNA, and omitting template from both RT and PCR reactions to ensure amplified product was not the result of laboratory contamination of reagents. For selected experiments, extracted RNA was also amplified by RT-PCR using primers specific for a 600 bp fragment of the human glyceraldehyde phosphate dehydrogenase (GAPDH) gene (Stratagene, La Jolla, CA). This is a "house-keeping" gene which is expressed by all human cells (HCT-8 are human cells) and its amplification serves as an internal control of the RNA extraction and RT-PCR procedures. The GAPDH amplicon should be present in all samples at the same signal strength regardless of the oocyst dose applied to the cells or the level of inactivation or reactivation of oocysts.

ANIMAL INFECTIVITY

A limited number of mouse infectivity assays were performed to support the cell culture-based data. Dams and neonatal CD-1/ICR mice (5- to 7-days old) were obtained from Charles River Laboratories, Inc. (Wilmington, MA) and neonates were inoculated by delivering a 10 µL aliquot of each dose of oocysts (doses ranged from 100 to 50,000 oocysts), suspended in sterile water, to the back of the throat using a calibrated pipette. Cages were labeled with date of infection, dose, number of pups infected and study code. Seven days after inoculation, formalin fixed, slide mounted sections of the terminal ileum (5 µm × 2–3 cm) were stained with hematoxylin and eosin (Korich et al., 2000) and infections were detected by microscopic examination. The ileum was chosen because it is the first region of the gut to be colonized by the parasite and harbors the majority of development stages (Current and Reese, 1986). Evidence of infection was defined as the observation of *C. parvum* parasite developmental stages in the microvilli of any of the prepared histological sections. Tissue blocks and labeled slides are archived for 5 years.

QUANTIFICATION OF INFECTIVITY AND INACTIVATION

Each challenge dose of either untreated oocysts or UV-exposed oocysts was inoculated into at least six wells of HCT-8 monolayers or 10 mice. All infectivity results were expressed as the percent of cell culture monolayers or animals that became infected at each oocyst dose. Dose response curves were constructed by fitting a least squares regression to a plot of oocyst challenge dose against a logistic transformation of the proportional infectivity (Korich et al., 2000; Rochelle et al., 2002). The response logit (*RL*) is defined as:

$$RL = Log_e \frac{(P/100)}{[1-(P/100)]}$$

where Log_e = the natural logarithm

P = the percentage of infected wells or mice

The ID_{50} values were calculated by solving the logistic regression for a logit value of zero. To determine the level of inactivation, dose response curves were constructed for untreated control oocysts and oocysts exposed to UV light and the level of inactivation was then determined by comparing the ID_{50} of exposed oocysts to the control, untreated oocysts. Inactivation was expressed as the logarithm (base 10) of the increase in ID_{50} following exposure to UV:

$$Inactivation = Log_{10}(ID_{50}^{C}) - Log_{10}(ID_{50}^{UV})$$

where ID_{50}^{C} = the 50% infectious dose for control oocysts

ID_{50}^{UV} = the 50% infectious dose for UV exposed oocysts

The logit function does not allow logarithmic transformations of 0 and 100% responses so any set of monolayers or cage of animals that generated 0 or 100% infection (P = 0% or 100%) was excluded from the logit response model. The logit dose response method is an accepted analytical tool for expressing parasite infectivity and has been used previously for inter-laboratory comparisons of *C. parvum* infectivity in mice (Korich et al., 2000) and for comparing cell culture with a mouse infectivity assay (Rochelle et al., 2002).

OOCYST REPAIR EXPERIMENTS

Following exposure to UV light, oocysts were either assayed immediately to determine the amount of inactivation or exposed to a variety of potential repair conditions prior to assaying for infectivity. Oocysts were incubated in phosphate buffered saline at 37°C in darkness for 10 – 30 min and in 0.5% sodium hypochlorite (10% bleach) or 0.75% sodium taurocholate at 37C for 10 – 20 min prior to inoculation of cell monolayers with and without a holding period of 24 h under fluorescent white light or in complete darkness at room temperature immediately following UV exposure. Cell monolayers were also incubated for up to five days following inoculation with UV-exposed oocysts instead of the usual 48 h. UV-exposed oocysts were also incubated at room temperature for 2 h, followed by incubation in 0.75% sodium taurocholate at 37°C for 15 min, exposure to fluorescent white light in a humidified incubator at room temperature for 2 h, and finally incubation at room temperature in darkness for 16 h prior to inoculation of cell monolayers. In addition, mice inoculated with UV exposed oocysts were maintained for 10 days prior to detecting infection instead of the usual 5 days.

Oocysts were exposed to photoreactivation conditions in a humidified full-spectrum light box that vented heat of convection from the daylight lamps, while assuring no volumetric sample loss to evaporation. Inside the light box, samples were maintained at room temperature. Dark repair experiment samples were placed inside a covered container to protect samples from any stray light.

DETECTION OF UV REPAIR GENES

Identification of DNA Repair Genes in the *C. parvum* Genome

Nucleotide excision repair genes in *C. parvum* were identified by a combination of data-mining existing nucleic acid sequences in various databases and physical sequencing of genes and gene fragments amplified from a variety of isolates. The *C. parvum* genomic database generated by the University of Minnesota *Cryptosporidium parvum* Genome (MCPG) sequencing project (www.ncbi.nlm.nih.gov/Microb_blast/credits/5807.html) was screened using BLAST searches (Atschul et al., 1997) with known sequences of DNA repair genes from a variety of organisms, including selected bacteria, *Arabidopsis thaliana*, *Homo sapiens*, and *Saccharomyces cerevisiae.*

Some genes were also identified by random sequencing of 200 clones of a genomic DNA library of the KSU-1 isolate (Khramtsov et al., 1995). The *C. parvum* database was also screened using sequences encoding photolyase enzymes (GenBank accession codes AAA34875, AAB04997, P40115, NP-469940), cyclobutane pyrimidine dimers (CPD)-specific DNA glycosylases (AF378536, AF128168), and UV-specific endonucleases (U70046, AJ243712, NP-005227, NP-015303). The MCPG sequencing project provides up to seven-fold genome coverage of the Iowa isolate of *C. parvum*. The sequences of UV repair genes that were identified in the MCPG database were also BLAST screened against the genomic database for an anthroponotic (genotype 1) isolate of *C. parvum* provided by the Virginia Commonwealth University and Tufts University School of Veterinary Medicine *Cryptosporidium parvum* Genome Sequencing Project accessible at www.parvum.mic.vcu.edu.

Complementation Assay to Assess Gene Function

Following identification of repair genes in the *C. parvum* genome database, PCR primers were designed allowing amplification of the various genes from genomic preparations of the KSU-1 isolate. The resulting amplicons were cloned into the pYES2.1 TOPO yeast expression plasmid vector (Invitrogen) utilizing the overhanging deoxyadenosine residues incorporated at the 3' end of each amplified strand by thermostable Taq polymerase (TA cloning). Following amplification with RAD gene specific primers, amplicons were mixed with the cloning vector, incubated at room temperature for five minutes, and then transformed into *E. coli* XL-1 Blue (Stratagene) using standard transformation protocols (Sambrook et al., 2001). The resulting transformant colonies were screened for the presence of RAD genes in the correct orientation by restriction digestion and sequenced utilizing priming sites incorporated into the vector by the DNA Sequencing Facility at the KSU Plant Pathology Department. Confirmed transformant colonies were grown in nutrient broth culture and plasmid DNA was extracted and purified using standard procedures (Sambrook et al., 2001). This plasmid DNA was then transformed into various mutant strains of *Saccharomyces cerevisiae* (auxotrophic for uracil, *ura3*, to allow selection of the pYES2.1 TOPO-RAD construct, and galactose-utilizing to induce expression) to generate recombinant yeast strains carrying *C. parvum* DNA repair genes. Transformants were selected on medium lacking uracil and containing galactose to induce the heterologous repair gene. These cells were exposed to UV radiation from a germicidal UV bulb (0.5 to 5 mJ/cm^2) and incubated for three days at 30°C. The irradiated plates were observed for colony growth to determine whether cross-species complementation had occurred.

Screening Isolates for the Presence of Repair Genes

To screen a wider variety of isolates for the presence of UV repair genes, PCR primers were designed with the aid of the LightCycler Probe Design software (Roche) to amplify specific regions from the RAD1, RAD2, and RAD3 genes that were identified in the *C. parvum* genome during this project (Table 2.3). BLAST homology screening of the primer sequences against the GenBank database at the National Center for Biotechnology Information (www.ncbi.nlm.nih.gov) demonstrated no significant homology with any sequences. Screening against the apicomplexan genomes database revealed identity with the respective *C. parvum* sequences. Since the only intended use for these primers was to amplify target regions from DNA extracted from purified oocyst preparations, more extensive empirical specificity screening was deemed unnecessary.

These primers were used to assess the occurrence of the selected repair genes in a variety of isolates. Genomic DNA was extracted from fresh oocysts of the Iowa isolate and 2% potassium dichromate preserved oocysts of the Moredun, Glasgow, Maine, and TAMU isolates. Oocysts stored in potassium dichromate were washed five times by repeated centrifugation at 10,000 rpm for 3 min and resuspension in PBS prior to extraction. Oocyst DNA was extracted using a previously described hexadecyltrimethylammonium bromide (CTAB)-proteinase K protocol (Tsai and Rochelle, 2001). The absorbance spectrum for DNA extracted from the Iowa isolate of *C. parvum* was measured using a scanning UV-VIS spectrophotometer (Shimadzu, UV2401PC). Comparison with the spectra for *S. cerevisiae* (yeast) DNA extracted using the same procedure and with purified plasmid DNA demonstrated that *C. parvum* DNA absorbed UV light as expected (Figure 2.4). This ruled out aberrant light absorbance as a cause of any unusual response to UV light or presence/absence of repair mechanisms. Amplification conditions for all three primer sets were 10 mM Tris-HCl, pH 8.3, 50 mM KCl, 3 mM $MgCl_2$, 0.2 mM of each dNTP, 0.25 μM of each primer (Table 2.3), 0.01 U/μL of UDG, and 0.025 U/μL of Platinum Taq polymerase. Annealing temperatures ranging from 52°C to 62°C were evaluated using a PTC-200 gradient thermal cycler (MJ Research) but there was no observed difference in the PCR signal strength. Therefore, an annealing temperature of 60°C was selected for all subsequent amplifications with these primers. The primers amplified 369-bp (base positions 1571 – 1939), 304-bp (positions 2406 – 2709), and 345-bp (positions 282 – 626) fragments from the CpRAD1, CpRAD2, ad CpRAD3 genes, respectively.

Following 40 cycles of amplification of DNA from each isolate, amplicons were purified by passage through Wizard PCR Prep columns following the manufacturer's instructions (Promega) and resuspended in 50 μL of sterile deionized water. The nucleic acid sequences of both strands of the RAD amplicons were determined by cycle sequencing (50 cycles) using both amplification primers as sequencing primers, ABI BigDye Terminator v3.1 sequencing kit, AmpliTaq DNA polymerase and an Applied Biosystems 3100 capillary analyzer by the Molecular Genetics Instrumentation Facilities at the University of Georgia (Athens, Georgia). A consensus sequence was then generated for each gene fragment from the forward and reverse sequences and the sequences analyzed using the DNASIS software package (Hitachi Software).

ASSESSMENT OF UV-INDUCED DNA DAMAGE

A variety of methods were used to assess the extent of UV-induced damage in oocyst nucleic acids, with varying levels of success. These methods included quantitative PCR, a Comet assay, and an anti-thymine dimer antibody. Since the Comet assay approach (based on single cell gel electrophoresis) did not produce useful results it is not described further. In addition, an endonuclease site sensitive assay (Oguma et al., 2001) was used to assess the extent of DNA damage (experiments conducted by Z. Bukhari at American Water, Belleville, IL). Both UV treated and untreated *E. coli* and *C. parvum* DNA preparations were digested with T4 endonuclease V (T4-pyrimidine dimer glycosylase; Trevigen, Gaithersburg, MD) for 30 min at 37°C (1 U per 250 ng of genomic DNA). The reactions were stopped using an alkaline loading dye and DNA was separated by electrophoresis on a 0.5% alkaline agarose gel in alkaline buffer containing 30 mM NaOH and I mM EDTA. However, these experiments were unable to replicate the work of Oguma et al. (2001) because the underlying premise of shifting DNA molecular weight distributions from high to low molecular weight following UV exposure could not be duplicated (Z. Bukhari, personal communication).

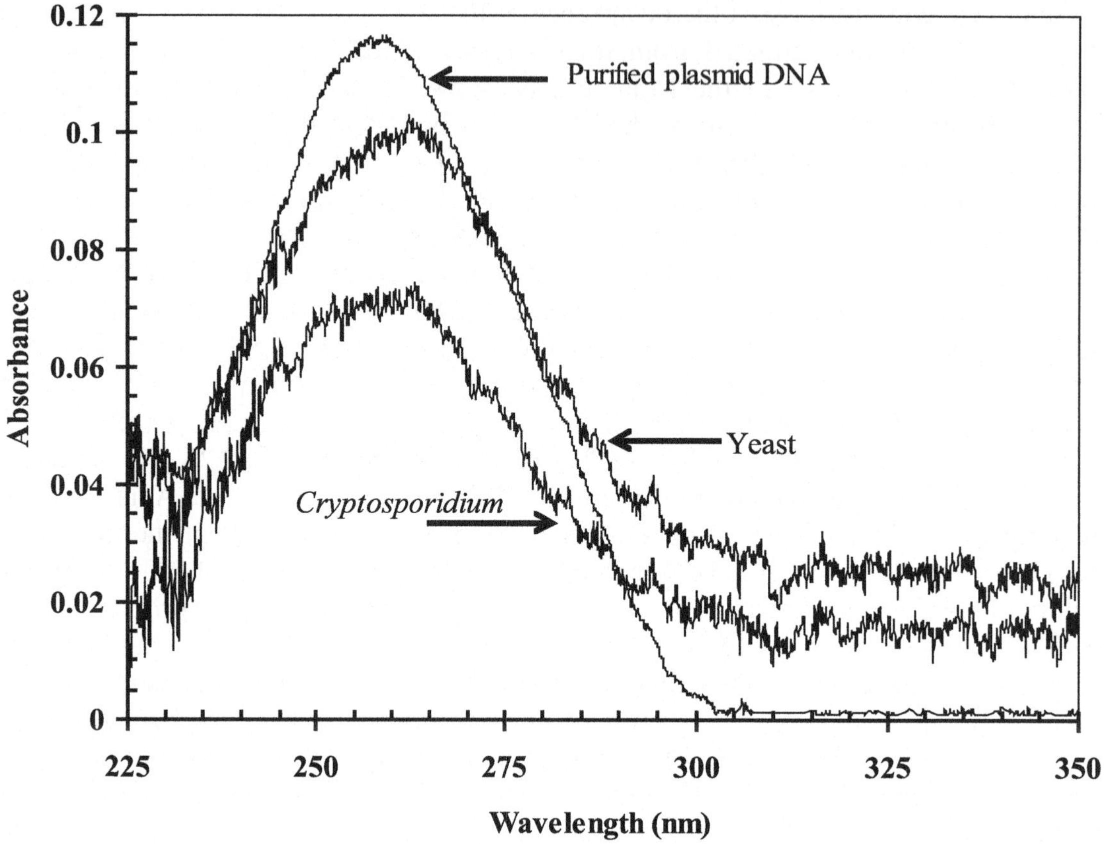

Figure 2.4 Absorbance spectra of genomic DNA extracted from *Cryptosporidium parvum* (Iowa isolate) and *Saccharomyces cerevisiae* (yeast) compared to the spectrum of a purified plasmid DNA preparation

Anti-Thymine Dimer Antibody Assay

The extent of UV-induced DNA damage was assessed through the use of a monoclonal antibody that recognized and bound specifically to cyclobutane thymine dimers (mouse antibody H3; Bio-Can). Control experiments demonstrated that this antibody did not bind to DNA that had not been exposed to UV and therefore did not contain thymine dimers.

Following exposure of Iowa isolate oocysts and yeast cells to increasing doses of medium pressure UV light, genomic DNA was extracted using a CTAB-proteinase K-phenol/chloroform procedure (Tsai and Rochelle, 2001). The DNA was digested with the enzyme *Hin*dIII (37°C, 3 h), separated by agarose gel electrophoresis, and blotted to a positively charged nylon membrane using standard techniques (Sambrook et al., 2001), without staining or photographing to avoid exposure of the extracted DNA to UV light from the transilluminator. The membrane was air-dried at room temperature, washed for 30 min in 0.1 M maleic acid, 0.15 M NaCl, pH 7.5 containing 4% blocking reagent (Roche), and then incubated at 37°C for 30 min in 0.1 M Tris-HCl, pH 7.5, 0.15 M NaCl, 2% blocking reagent containing 2 µg/mL of unlabeled H3 anti-thymine dimer antibody. This was followed by 30 min incubation at 37°C in 0.1 M Tris-HCl, pH 7.5, 0.15 M NaCl, 2% blocking reagent containing alkaline phosphatase-labeled goat anti-mouse IgG_1 (SouthernBiotech, Birmingham, AL). Following three 15 min washes in 0.1 M maleic acid, 0.15 M NaCl, pH 7.5, containing 0.3% Tween-20, the bound antibody was visualized by incubation in a chemiluminescent substrate (CPD Star, Roche) diluted 1 in 100 in 0.1 M Tris, 0.1 M NaCl, 50 mM $MgCl_2$, pH 9.5 at room temperature for 15 min, and exposure of the membrane to photographic film for 5 sec. The film was developed using standard procedures and signal intensity of the digitized image was measured using densitometric software (ProQuanta, DNA ProScan, Inc., Nashville, TN).

Quantitative PCR

The extent of UV-induced DNA damage was also measured using a quantitative PCR method. Quantitative sequence detection (QSD), a real-time quantitative PCR technique targeting the *C. parvum* hsp70 gene was used to directly quantify UV-induced damage to oocyst DNA.

Exposure of Oocysts to Low and Medium Pressure UV

A range-finding trial was performed using a Rayox medium-pressure UV collimated beam unit (Model PS1-1-120, Calgon Carbon Corp., Pittsburgh, PA) and a radiometer equipped with a SED240 UV detector calibrated by the manufacturer to standards of the National Institute of Standards and Technology (Model IL1400A, International Light Inc., Newburyport, MA). Additional trials were performed using a low-pressure bench-scale monochromatic 254 nm low-pressure UV system with integrated calibrated radiometer (Stratalinker 1800 UV crosslinker; (Stratagene). For each experiment, a single suspension of oocysts at a concentration of 6×10^4/mL was used. Oocysts were exposed cumulatively to 0, 5, 10, 20, 40, 60, 100 mJ/cm^2 UV dosages. Immediately after each exposure, suspensions were thoroughly mixed and a 100 μL aliquot of exposed oocysts was transferred to a 0.65 mL microfuge tube containing 100 μL of a 1:1 Chelex and 10 mM Tris-1 mM EDTA (TE) slurry. The same suspension was again mixed well and exposed again to achieve the next dosage. Samples were immediately frozen at -80°C until DNA extraction and quantitative PCR. In addition, oocysts exposed to medium pressure UV at MWD were immediately frozen at -80°C (without addition of Chelex/TE) and shipped on dry ice to EP AREC for processing and analysis.

DNA Extraction From Oocysts

Oocyst DNA was prepared using a rapid and simple freeze-thaw extraction procedure. Biotechnology grade Chelex 100 resin (Bio-Rad, Hercules, CA) was equilibrated overnight in a 1:1 volume-to-volume ratio of TE (pH 8.0) to produce a slurry. An equal volume of prepared Chelex resin slurry was added to each sample using a micropipettor fitted with large-bore (genomic) tips. For each sample, the Chelex slurry was thoroughly resuspended and aliquots quickly withdrawn from the reagent tube to prevent settling of the resin and maintain a 1:1 volume-to-volume ratio of resin and TE buffer. Oocysts were lysed by eight 1 min cycles of freezing in liquid nitrogen and thawing at 98°C. Total sample volumes were 200 μL or less to ensure complete freezing and thawing during 1 min cycles.

Quantitative Sequence Detection (QSD)

Two different QSD (quantitative PCR) methods were performed using an Applied Biosystems 7700 or 7000 Sequence Detector (Foster City, CA). The first QSD method employed PCR primers and a TaqMan probe specific for the *C. parvum* hsp70 gene (Di Giovanni et al., 2002). The hsp70 primer and probe sequences are provided in Table 2.3. Each 50 μL QSD reaction contained 5.0 μL of × TaqMan A amplification buffer (Applied Biosystems), 3.0 mM $MgCl_2$, 200 μM each of dATP, dCTP, and dGTP (Amersham Biosciences, Piscataway, New Jersey), 800 μM of dUTP (Amersham Biosciences), 200 nM of each forward and reverse CPHSPT2 primer and CPHSP2P2 TaqMan probe (Applied Biosystems), 2.5 μL of 30 mg/mL bovine serum albumin (BSA; Sigma), 2.5 U AmpliTaq Go DNA polymerase (Applied Biosystems), and 1 U of UDG (New England Biolabs, Beverly, Massachusetts). Amplification conditions were: UDG digestion at 50°C for 10 min; initial denaturation at 95°C for 10 min; 45 cycles of denaturation at 95°C for 30 sec and annealing at 60°C for 1 min; followed by a final extension at 72°C for 10 min (when gel based confirmation was desired).

A novel QSD method was used to evaluate low dose (<20 mJ/cm) UV damage. This method used the GPD62 and GPD63 random amplified polymorphic DNA (RAPD) primers described previously (Table 2.3; Deng and Cliver, 1998) and SYBR Green I quantitative PCR detection chemistry (Applied Biosystems). Each 50 μL QSD mixture contained 5.0 μL of 10× SYBR Green amplification buffer (Applied Biosystems), 4.0 mM $MgCl_2$, 200 μM of each dATP, dTTP, dCTP, and dGTP (Amersham Biosciences), 1 μM of each RAPD primer, 2.5 μL of 30 mg/mL BSA and 2.5 U AmpliTaq G DNA polymerase (Applied Biosystems). Amplification conditions were: initial denaturation at 95°C for 10 min; 45 cycles of denaturation at 94°C for 30 sec and annealing at 40°C for 1 min; extension at 72°C for 2 min; amplicon detection at 76°C for 30 sec; followed by a final extension at 72°C for 7 min (when gel based confirmation was desired).

For QSD, 5 μL aliquots of Chelex-treated oocyst DNA were used as template. Molecular grade water was used to prepare no template negative controls. UV inhibitor controls consisted of molecular grade water exposed to 1000 mJ/cm^2 low pressure UV, used to prepare a PCR master mix, then seeded with Chelex DNA. For the hsp70 QSD, a standard curve and equation for quantitation was generated based on sample cycle threshold (Ct) values (the fractional PCR cycle number at which amplification was detected above background) plotted against the amount of oocyst DNA. Experimental data were analyzed and expressed as percentages of experimental controls. For RAPD QSD data, only Ct values of samples were analyzed.

CHAPTER 3
RESULTS AND DISCUSSION

This project used a combination of database mining, empirical experimentation, infectivity assays, and molecular assays to determine whether *C. parvum* oocysts can repair UV-induced DNA damage and regain infectivity following exposure to UV light. Infectivity and inactivation were assessed in most experiments by using an in-vitro cell culture-based approach. There are a variety of in-vitro surrogate methods that can be used for determining the viability of *C. parvum* oocysts but it has been demonstrated that these methods do not always correlate with infectivity assays (Bukhari et al., 2000). In fact, the initial studies of UV disinfection assessed inactivation by incorporation of vital dyes and in-vitro excystation and reported that a dose of 8,748 mJ/cm^2 was necessary to achieve >2-$\log_{10}$ (99%) inactivation (Campbell et al., 1995). Consequently, it is essential that assays used to measure disinfection determine the ability of oocysts to cause infection, rather than simply detect some aspect of metabolism or the presence of particular molecules. A recent study demonstrated that in-vitro cell culture combined with RT-PCR was equivalent to a standard mouse assay for measuring infectivity of untreated fresh oocysts (Rochelle et al., 2002). Therefore, this investigation employed a cell culture-based approach for measuring oocyst infectivity and inactivation.

MEASURING INFECTIVITY OF *C. PARVUM* OOCYSTS

Infectivity was measured by inoculating oocysts onto monolayers of HCT-8 cells, incubating at 36°C for 48 h, extracting total cellular RNA (host cell and parasite), and detecting infection by reverse transcriptase-PCR (RT-PCR) with primers specific for a *C. parvum* heat shock protein gene (Rochelle et al., 1997, 2002). Amplification products were visualized by standard agarose gel electrophoresis and staining with ethidium bromide (Figure 3.1). Infectivity was expressed as a logit transformation of the proportion of cell culture wells that became infected at each oocyst challenge dose. The logit dose response method linearizes infectivity data (it excludes 0% and 100% infectivity) and is accepted as an analytical tool for expressing parasite infectivity. It has been used previously for inter-laboratory comparisons of *C. parvum* infectivity in mice (Korich et al., 2000) and for comparisons between cell culture and mouse assays (Rochelle et al., 2002). Extracted RNA was also amplified by RT-PCR using primers specific for a 600-bp fragment of the human glyceraldehyde phosphate dehydrogenase (GAPDH) housekeeping gene in selected experiments. This amplicon should be present in all cell monolayers at the same signal strength regardless of the density of oocysts inoculated onto the cells (Figure 3.1).

Multiple genotype 2 isolates of *C. parvum* were used throughout the project and each inactivation experiment included untreated oocyst controls. Dose response curves were generated using the cell culture/RT-PCR method for untreated oocysts of all of the isolates that were used for inactivation studies. These dose response curves demonstrated very good congruence with previous cell culture-based evaluation of the infectivity of these isolates (Rochelle et al., 2002). Aggregate data for the Iowa isolate generated over a 15-month period for this project produced an ID_{50} of 73 oocysts (LR = [2.7965 × oocyst dose] – 5.2125; Figure 3.2). The average age of the oocysts at the time of infection was 32 days post-shedding (range, 9 – 104 days). In a previous study utilizing the same cell culture/RT-PCR methods, the ID_{50} for untreated oocysts of the Iowa isolate was 77 (95% confidence interval, 52 – 154; Rochelle et al., 2002) with a similar dose

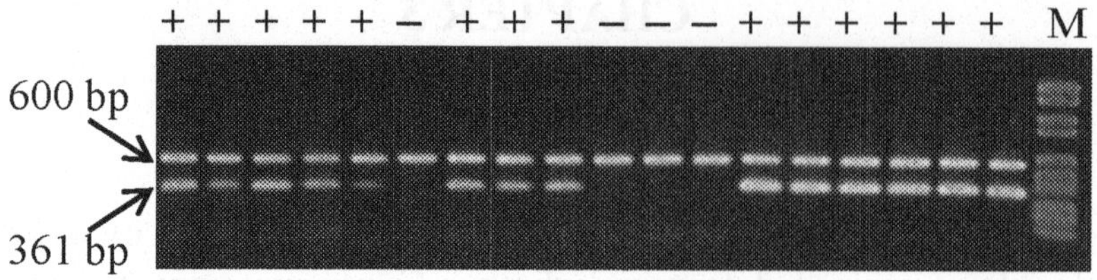

Figure 3.1 Example of the detection of infected cells by RT-PCR followed by gel electrophoresis. Separate amplification reactions were performed on total RNA extracted from cell monolayers with primers specific for the *C. parvum* hsp70 gene (361 bp) and the GAPDH internal control (600 bp). The resulting RT-PCR products were separated by electrophoresis on a single gel. Only monolayers that developed infection generated a *C. parvum* hsp70-specific amplicon (+). Monolayers that did not become infected generated the GAPDH amplicon (indicating that the extraction and RT-PCR procedures were working) but did not yield a *C. parvum*-specific product (–). A molecular size standard (M) was included to confirm amplicon size.

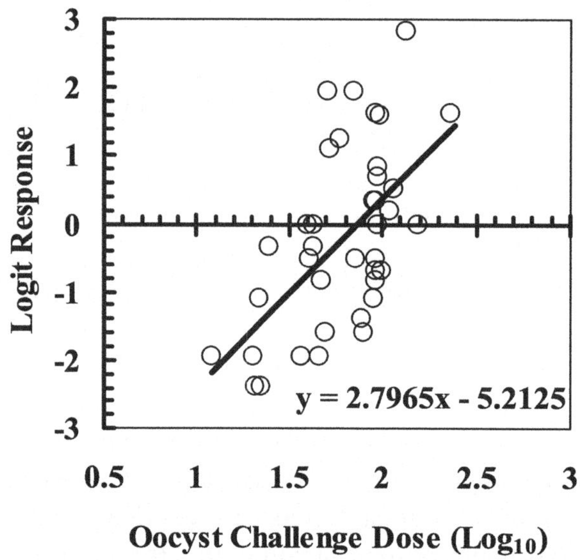

Figure 3.2 Dose response curve for the Iowa isolate of *C. parvum* in HCT-8 cell culture. The data was derived from all control oocysts that were not exposed to UV light throughout this project. The average age of the oocysts at the time of infection was 32 days from the day of shedding by the propagation animal. The ID_{50} (73 oocysts) was obtained by solving the regression equation for *Y = 0*.

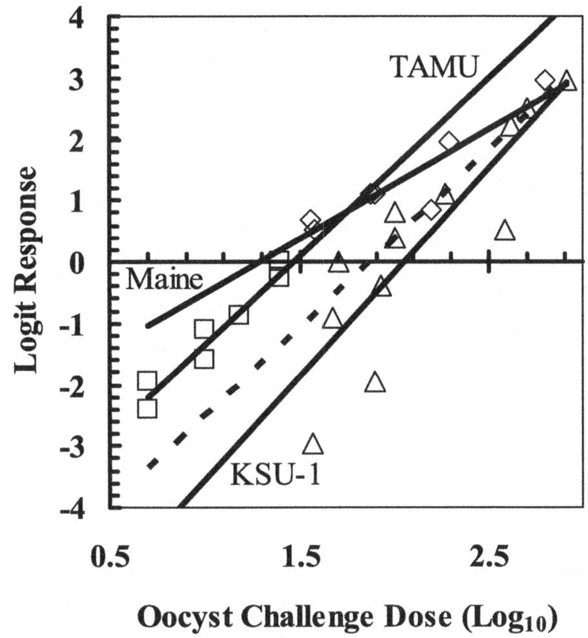

Figure 3.3 Dose response curves for untreated oocysts of the KSU-1 (△), Maine (◇), and TAMU (□) isolates of *C. parvum* compared to the Iowa regression line from Figure 3.2 (dashed line)

response regression. The level of variability in the data was also similar. This previous study also demonstrated that the HCT-8 cell culture/RT-PCR method was equivalent to the standard CD-1 mouse assay for measuring infectivity of fresh, untreated oocysts. The consistency of dose response data and ID_{50} values generated in different projects over relatively long periods of time, and by different laboratory personnel, confirms the robustness of in-vitro cell culture combined with RT-PCR for measuring the infectivity of *C. parvum.*

Dose response curves were also generated for untreated oocysts of the other isolates used in this investigation (Figure 3.3). The average ID_{50} values for the TAMU, KSU-1 and Maine isolates based on aggregate data from multiple experiments were 29, 111, and 20 oocysts, respectively. The TAMU isolate has been reported to have an ID_{50} of 9 oocysts in human feeding trials (Okhuysen et al., 1999) compared to 132 for the Iowa isolate (DuPont et al., 1995). However, the variability inherent in all infectivity assays, whether conducted in humans, animal surrogates, or cell culture, makes it difficult to draw conclusions on the comparative infectivity of various isolates. In previous study, we demonstrated that the Maine isolate had an ID_{50} of 21 in CD-1 mice and 86 in HCT-8 cell culture (Rochelle et al., 2002). In the present investigation the ID_{50} was 20 oocysts, close to the value of 21 previously reported for mice. The purpose of generating dose response curves for untreated, control oocysts for each of the isolates used in this investigation was to confirm that the oocysts and the cell culture-based infectivity assay were performing as expected. This allowed us to rule out abnormal oocyst infectivity as an explanation for any erroneous UV-inactivation and/or reactivation results that occurred.

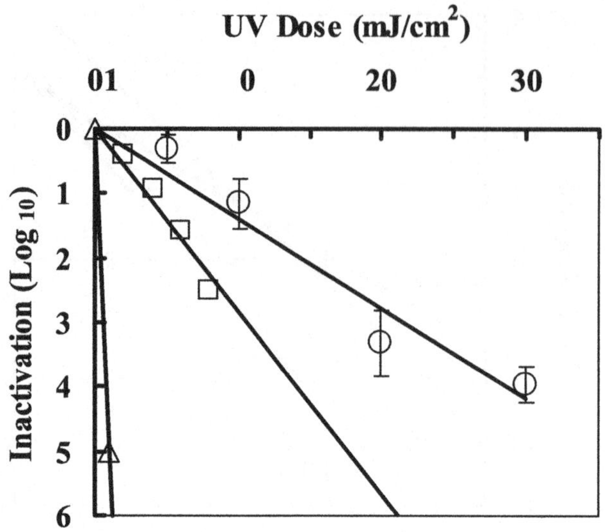

Figure 3.4 Disinfection of bacteria by UV light. ***Bacillus subtilis*** **spores (○), a repair-competent strain of** ***Escherichia coli*** **(ATCC 15597, □) and a UV-sensitive mutant of** ***E. coli*** **(ATCC 23229, △) were exposed to medium-pressure UV light. The error bars on the** ***B. subtilis*** **data represent the standard deviation of multiple replicates.**

INACTIVATION OF BACTERIA AND YEAST BY UV LIGHT

UV disinfection of *Bacillus subtilis* and *Escherichia coli* was measured to ensure that the UV equipment used for this investigation was generating the expected levels of inactivation (Figure 3.4). UV dose response curves were also generated for three strains of *Saccharomyces cerevisiae* (Figure 3.5). Also, these control organisms were used to ensure that the experimental equipment or set-up did not prevent UV-repair from taking place in repair-competent organisms. The most extreme example of UV inactivation was represented by a UV-sensitive mutant of *E. coli*; >5-$\log_{10}$ (>99.999%) inactivation with UV doses <2 mJ/cm^2. In contrast, a UV-resistant strain of *E. coli* demonstrated 2-$\log_{10}$ (99%) inactivation at a UV dose of approximately 6.5 mJ/cm^2. As expected, *B. subtilis* spores were the most resistant form of bacteria tested; a dose of approximately 27 mJ/cm^2 was necessary to achieve 4-$\log_{10}$ (99.99%) inactivation (Figure 3.4).

Inactivation curves for two UV sensitive strains of *S. cerevisiae* that had mutations in one or more of their RAD UV-repair genes demonstrated 5-$\log_{10}$ (99.999%) inactivation at UV dosages of 3.5 and 7.7 mJ/cm^2 (Figure 3.5). In contrast, the inactivation curve for a UV repair competent strain demonstrated only a 1-$\log_{10}$ (90%) reduction at 6 mJ/cm^2. When exposed to white light for 2 h following UV exposure, cells of this latter strain displayed up to 0.91-$\log_{10}$ less inactivation than cells that were enumerated immediately following UV exposure. This result demonstrates the repair capacity of this particular strain of yeast.

The bacterial and yeast results combined demonstrated that: 1) Expected levels of inactivation were achieved with a range of relatively simple prokaryotic and eukaryotic microorganisms; 2) There is considerable intra-species variability in the response of microorganisms to UV light; and 3) Repair-competent microorganisms are not inhibited from repairing UV-induced damage by using the experimental equipment intended for this project.

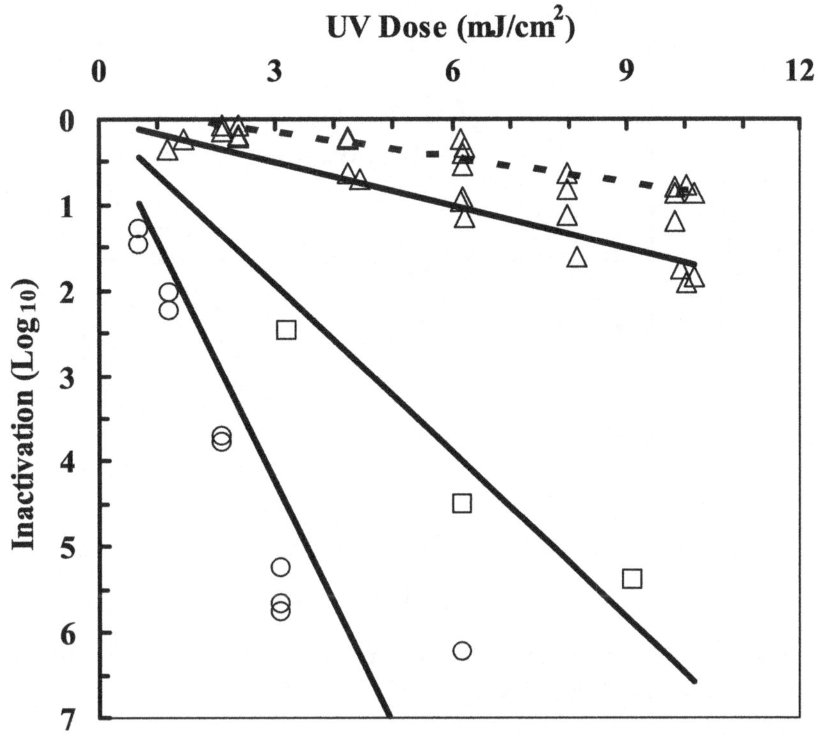

Figure 3.5 Inactivation of *Saccharomyces cerevisiae* by UV light. Cells of *S. cerevisiae* were plated on TSA plates immediately following exposure to UV light (——) or following incubation for 2 h in a light box to stimulate photoreactivation (- - -). The strains used were UV-sensitive mutants that were deficient in one or more of the RAD repair genes (○ and □) and a repair competent strain (△).

UV INACTIVATION OF *C. PARVUM* OOCYSTS

For *C. parvum* disinfection and repair studies, dose response curves were constructed for untreated oocysts and oocysts exposed to UV light and the level of inactivation was then determined by comparing the ID_{50} ($Y = 0$ in regression equation) of exposed oocysts to the control, untreated oocysts (Figure 3.6). In this particular example, the ID_{50} for control oocysts of the KSU-1 isolate was 53.5 (○) compared to 2.1×10^5 oocysts for UV-exposed oocysts (□). Therefore, the level of inactivation was 3.6-$\log_{10}$ (99.98%). In some instances, UV-exposed oocysts only generated infections at a single challenge dose of oocysts so there was insufficient data to generate a "treated" dose response curve. In these cases, the amount of inactivation was calculated by solving the control dose response regression for the proportional infectivity (logit transformed) generated by the available oocyst doses. Oocysts of five zoonotic isolates of *C. parvum* (Iowa, Moredun, Maine, TAMU, and KSU-1) were exposed to medium-pressure and low-pressure UV light using a bench-scale Calgon Carbon collimated beam instrument.

Oocysts of the Iowa isolate were 14 – 29 days post-shedding at the time of exposure to UV light and generated a linear inactivation curve in the range 0 – 8 mJ/cm^2 (Figure 3.7). Data from both low-pressure and medium-pressure UV lamps were combined since there was no difference

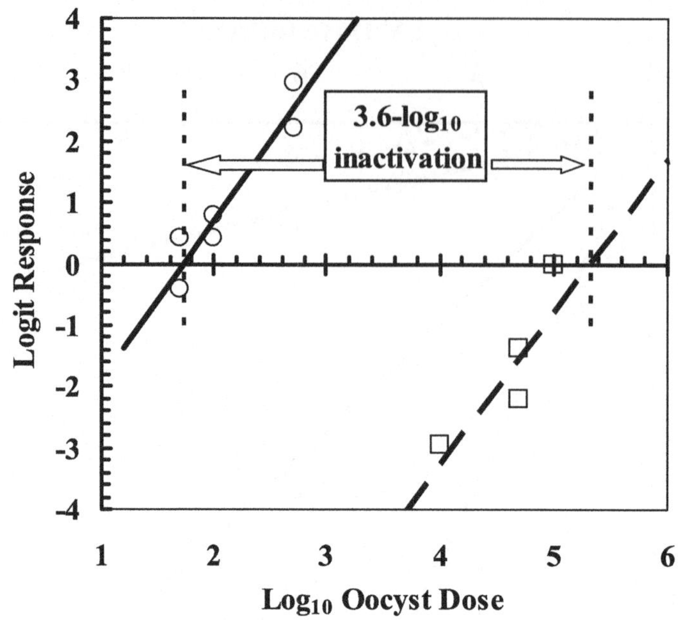

Figure 3.6 Measuring inactivation by comparing the dose response curves for non-exposed and UV-exposed oocysts of the KSU-1 isolate of *C. parvum*

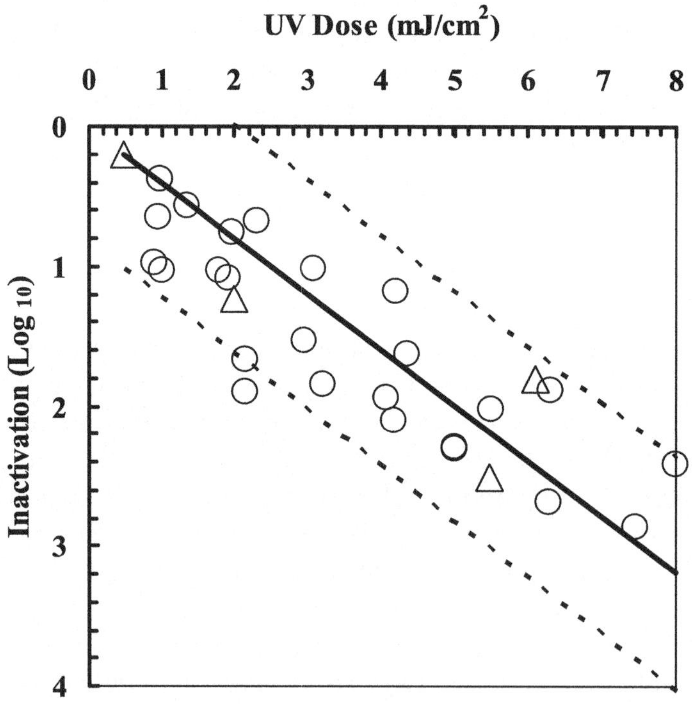

Figure 3.7 Inactivation of the Iowa isolate of *C. parvum* by UV light. The dashed lines represent the 90% prediction limits of the regression (R^2 = 0.60 for regression forced through origin and 0.74 for unforced regression). The data represent actual inactivation values determined by comparing ID_{50} values of control and exposed oocysts (○) or inactivation levels determined by solving the control dose response regression for a single infectivity value (△).

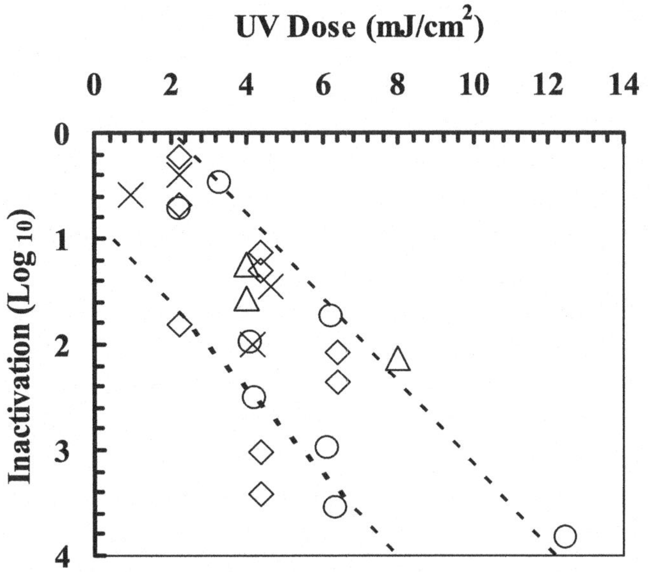

Figure 3.8 Inactivation of a variety of genotype 2 isolates of *C. parvum* by UV light. The dashed lines represent the 90% prediction limits calculated from the Iowa inactivation data. Inactivation was measured for the KSU-1 (○), TAMU (△), Moredun (×), and Maine (◇) isolates, following exposure to medium-pressure or low-pressure UV light.

in the amount of inactivation obtained with each type of lamp. The relationship between UV dose and inactivation was described by the linear regression equation:

$$Log_{10}(Inactivation) = 0.3982 \times UV\,dose$$

Therefore, 1-$\log_{10}$ (90%), 2-$\log_{10}$ (99%), and 3-$\log_{10}$ (99.9%) inactivation of the Iowa isolate was achieved with UV dosages of 2.5, 5, and 7.5 mJ/cm^2, respectively. These figures show good agreement with an earlier UV disinfection study that used cell culture and RT-PCR in a less replicated, semi-quantitative manner rather than the heavily replicated presence/absence approach used in this project. The previous study demonstrated 1-$\log_{10}$ (90%), 2-$\log_{10}$ (99%), and 3-$\log_{10}$ (99.9%) inactivation at medium-pressure UV dosages of 3, 6, and 9 mJ/cm^2 (Mofidi et al., 2001). When the lower 90% prediction limit (appears as the upper dashed line in Figure 3.7 because of the inverted Y-axis) of the Iowa data in the current project was used as a conservative measure of inactivation, which is more likely to be used for treatment plant design purposes, the average inactivation levels were 1-$\log_{10}$ (90%) and 2-$\log_{10}$ (99%) at UV dosages of 4.6 and 7.1 mJ/cm^2, respectively. In comparison, when the lower 95% prediction limit of medium-pressure and pulsed UV data combined was used in the in the earlier study as the conservative measure of disinfection, the average inactivation levels of 1-$\log_{10}$ (90%) and 2-$\log_{10}$ (99%) were achieved with polychromatic UV dosages of 7.5 and 11 mJ/cm^2, respectively (Mofidi et al., 2001).

For the other isolates used in this investigation, the results were similar to the Iowa isolate with a UV dosage of 6 mJ/cm^2 resulting in inactivation levels of 2.2-$\log_{10}$ (99.4%) for the Moredun isolate, 2.3-$\log_{10}$ (99.5%) for KSU-1, 1.8-$\log_{10}$ (98.2%) for the TAMU isolate, and 2.8-$\log_{10}$ (99.8%) for the Maine isolate (Figure 3.8). The slopes of the inactivation curves for the KSU-1 and Moredun isolates were not significantly different from the Iowa isolate indicating that all three isolates responded to UV in the same way. Although the Maine isolate appeared to be more sensitive and the TAMU isolate less sensitive compared to the Iowa isolate, the limited amount of data generated for these isolates and the relatively large variability in the inactivation data precluded assigning significance to these observations. Forced regressions of the inactivation data for each of the additional isolates fell within the 90% prediction limits derived from the Iowa inactivation data with the exception of TAMU, which crossed the lower 90% boundary at UV doses greater than 8 mJ/cm^2.

There is little published information on the response of these additional isolates to UV light so the significance of this work is the demonstration that various isolates respond similarly, with the same level of sensitivity, to UV light exposure. An earlier study demonstrated >4-$\log_{10}$ (>99.99%) inactivation of the Moredun isolate by 10 mJ/cm^2 of low-pressure UV light (Clancy et al., 2001). Overall, these inactivation results show a high level of consistency between different isolates and between different studies using similar methods. This indicates that UV inactivation results obtained with one isolate of *C. parvum* may be safely extrapolated to other isolates. However, genotype 1 *C. parvum* isolates (now named *C. hominis*) represent an exception to this conclusion, at least for the time being. No disinfection experiments have yet been conducted with this organism so there is no information on its response to UV light. *C. hominis* can not be easily propagated in standard animal models so it is difficult to obtain sufficient fresh oocysts to conduct disinfection experiments. Since it is now recognized as a separate species (Morgan-Ryan et al., 2002) and there are demonstrated biological, physiological and genetic differences between the two organisms, *C. parvum*-derived inactivation data cannot be extrapolated to *C. hominis*.

A limited number of experiments were also performed for this project using a CD-1 mouse assay to measure inactivation. Combined data for the Iowa, KSU-1, Moredun, and TAMU isolates demonstrated an average inactivation of 2-$\log_{10}$ (99%) with a UV dose of 5.6 mJ/cm^2 (Figure 3.9). This result shows good agreement with previously published inactivation data (Bukhari et al., 1999; Clancy et al., 2000; Craik et al., 2001; Hanes et al., 2002; Hargy et al., 2000). Although there is considerable scatter within the mouse-derived UV inactivation data (this project and published literature combined), most of the data fell within the 90% prediction boundaries of the cell culture-derived inactivation data for the Iowa isolate (Figure 3.7). This demonstrates the general congruence of mouse and cell culture assays for measuring UV disinfection of *C. parvum*. The overall agreement in UV inactivation data generated by various laboratories using different assay methods and a variety of isolates provides a strong measure of confidence to the conclusion that UV efficiently inactivates *C. parvum* (Table 3.1).

The majority of the UV inactivation data generated for this project was obtained using a medium-pressure collimated beam apparatus. However, approximately 20% of the data was derived from exposure to a low-pressure lamp. A comparison of the low-pressure and medium-pressure inactivation data is presented in Figure 3.10. Since similar levels of inactivation were obtained for all isolates of *C. parvum*, and there was agreement between the cell culture and mouse-derived inactivation results, all of the data was combined for this comparison. Although the regression lines indicate a slight difference in the levels of inactivation between the two types of lamp, the difference was not significant, as evidenced by the overlapping 95% confidence intervals of the respective regressions. Therefore, there was no difference in *C. parvum* inactivation between low- and medium-pressure UV light. This finding agrees with other studies of UV inactivation of *C. parvum* that used mouse assays (Clancy et al., 2000; Craik et al., 2001) and cell culture to measure infectivity and inactivation (Linden and Mofidi, 2003).

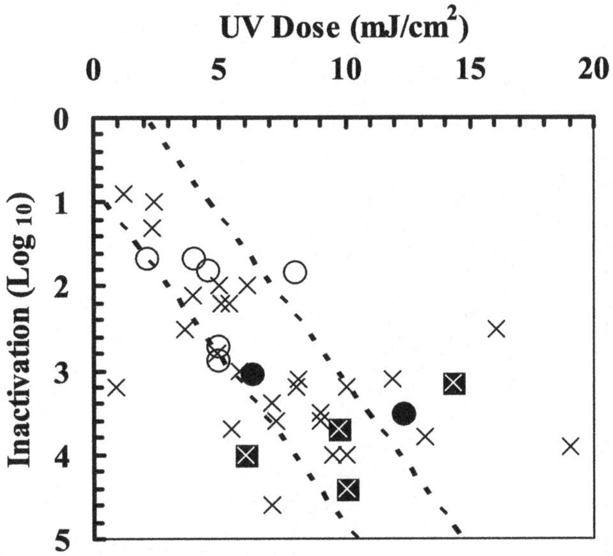

Figure 3.9 Inactivation of the Iowa, KSU-1, Moredun, and TAMU isolates of *C. parvum* by UV, measured using the CD-1 mouse assay (○). Filled symbols represent minimum inactivation levels because no mice developed infection with any challenge dose of oocysts. The dashed lines are the 90% prediction limits for cell culture-based Iowa inactivation data. Data from published literature using mouse assays (×, mostly neonatal CD-1 but also gamma interferon knockout and infant BALB/cByJ) are also plotted (Bukhari et al., 1999; Clancy et al., 2000; Craik et al., 2001; Hanes et al., 2002; Hargy et al., 2000).

Table 3.1
Summary of UV dose requirements for inactivation of *C. parvum*

	Necessary UV dose (mJ/cm^2) based on:				
Log_{10} inactivation	This project Iowa cell culture	This project all isolates cell culture	Mofidi et al., 2001, cell culture	This project all isolates mice	This project and literature combined mice
1	2.51	2.53	3.16	2.80	2.63
2	5.02	5.06	6.32	5.60	5.26
3	7.53	7.59	9.48	8.40	7.89

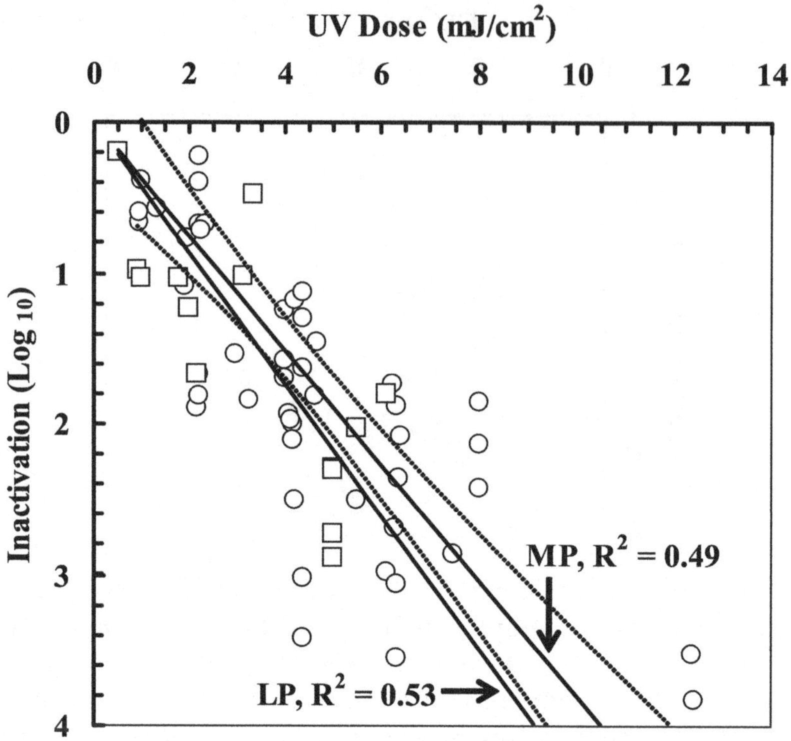

Figure 3.10 Comparison of medium-pressure (○) and low-pressure UV (□) for inactivation of *Cryptosporidium parvum*. The dashed lines are the 95% confidence limits of the medium-pressure regression data. Low-pressure 95% confidence limits were omitted for clarity but the interval was approximately the same size as that for the medium-pressure data. Inactivation data was combined from multiple isolates obtained using HCT-8 cell culture and the CD-1 mouse assay. R^2 values apply to regression lines forced through the origin.

POST-UV REACTIVATION STUDIES

Although *C. parvum* is readily inactivated by UV light, it is not known whether oocysts are killed by UV light or merely damaged such that they cannot immediately initiate infection. Earlier studies of UV exposure using surrogate measures of viability such as in-vitro excystation and uptake of vital dyes indicated that much higher doses of UV (up to 8,700 mJ/cm^2) were necessary to achieve 2-log_{10} (99%) inactivation (Campbell et al., 1995). Thus, UV-exposed oocysts appear able to excyst but not to infect. Consequently, if oocysts are sub-lethally damaged by UV they may have the potential to self-repair and regain infectivity. The ability of *C. parvum* oocysts to repair UV-induced DNA damage and reactivate would limit the practicality of UV disinfection and have serious implications for the use of UV as a disinfectant in the water industry. To overcome the oocyst repair function, it may be necessary to greatly increase applied UV doses or the concentration of post-UV chemical disinfectants, thus increasing the cost of treatment. Therefore, UV-exposed oocysts were incubated under a variety of potential repair conditions to determine whether they could regain their pre-exposure level of infectivity.

The primary UV-repair mechanisms are photoreactivation (light repair), base excision repair, and nucleotide excision repair. Photoreactivation is a light dependent process mediated by a photolyase enzyme and has been demonstrated in a wide range of organisms. The photolyase enzyme first binds to the UV-damaged DNA in darkness and is then activated by exposure to visible light (400 – 600 nm) which reduces FAD (one of two chromophores contained in photolyase). Once activated and upon exposure to long-wave UV or short-wave visible light (300 – 500 nm), the second chromophore transfers energy to $FADH^-$ and the pyrimidine dimer is split through a process of electron donation (Yasui and Eker, 1998), reverting the DNA to its original pre-exposure state. Experimental investigations of photoreaction typically involve exposing UV-irradiated cells to a combination of dark and light conditions for various periods of time. Base excision repair, mediated by pyrimidine dimer DNA glycosylase, is not widely distributed in nature. Nucleotide excision repair (NER) is a ubiquitous repair process in which damaged DNA is recognized by a multi-protein complex that makes dual incisions in the strand of the DNA containing the damaged nucleotide(s), resulting in removal of a fragment of DNA. The minimum set of proteins or protein complexes required for damage recognition and the dual cleavage reaction is approximately fifteen in eukaryotes (see Figure 1.2). Cleavage is followed by displacement of the damaged fragment and filling in of the gapped region by DNA polymerase. Members of the NER complex include proteins uniquely involved in NER and those that have additional functions in routine cellular physiology such as transcription. Since NER is a complex process involving the interaction of many proteins, the process is only likely to occur in actively metabolizing cells.

There are a variety of approaches that can be used to detect repair in organisms following exposure to UV light. For organisms such as yeast and bacteria, UV-exposed cells are typically held in dark and/or light conditions for 2 – 48 h prior to plating on culture media. These conditions allow photoreactivation or other repair mechanisms to occur prior to the cells entering their growth cycle. This approach can also be used for *C. parvum* oocysts, allowing time for oocysts to repair prior to inoculation onto cell monolayers or into mice. No evidence of reactivation was reported when oocysts were incubated at 25°C and 37°C for up to 4 h in under both light and dark conditions following exposure to UV dosages of 1 – 10 mJ/cm^2 (Shin et al., 2001). Incubating UV-exposed oocysts (1 – 4 mJ/cm^2) in 10% bleach and subsequent holding at 5°C and 25°C for up to 72 h in light and dark also failed to reactivate oocysts (Zimmer et al., 2001). Also, incubation at 20°C for up to 2 h in light and 24 h in the dark did not allow UV-exposed oocysts to reactivate, although it was reported that UV-exposed oocysts had the ability to repair thymine dimers based on the results of an endonuclease sensitive site assay (Oguma et al., 2001). Craik et al. (2001) did not investigate post-UV oocyst repair but UV-resistance was hypothesized as a possible explanation for an inactivation "tailing" effect. These previous studies investigated the potential for reactivation by treating oocysts in various ways following exposure to UV light. The oocyst stage of the parasite is only likely to reactivate if the cellular machinery inside the oocysts maintains a level of metabolism sufficient to allow repair-enzymes to operate. Dormant oocysts that are essentially in stasis, awaiting ingestion by a suitable animal host that provides an excystation and infection-friendly environment, are unlikely to undergo repair.

Since it is unlikely that the dormant oocyst stage of the parasite will be able to undergo repair and reactivation, the focus of the cell culture-based experiments in this project was on activated oocysts, released sporozoites, and the later stages of the life cycle. Excystation is the organism's response to the oocyst finding itself in favorable conditions for reproduction and may be induced by a variety of stimuli in vitro, such as acidic pH, incubation at 37°C, exposure to bile salts, or oxidants. Since the oocyst stage is essentially dormant, excystation must be able to occur with either no or a very low level of metabolism. It is known that oocysts that have been exposed to very high doses of UV can still excyst, releasing their sporozoites (Campbell et al., 1995). If

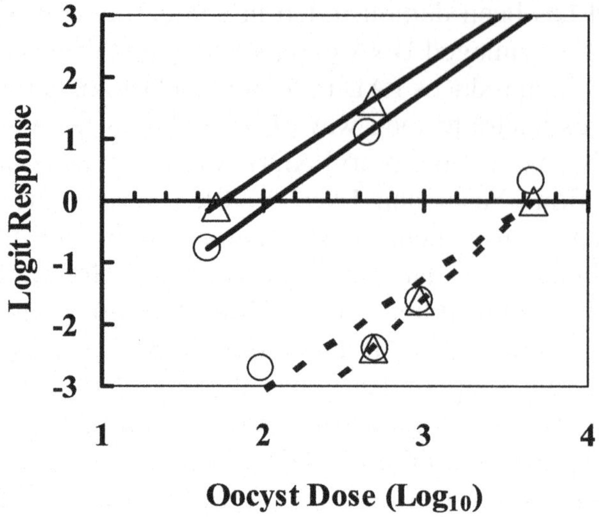

Figure 3.11 Example dose response curves for UV exposed Iowa isolate oocysts (2.15 mJ/cm², dashed lines) and untreated control oocysts (0 mJ/cm², solid lines). Oocysts were inoculated onto HCT-8 cells immediately following UV exposure (○) or incubated in 0.75% sodium taurocholate at 37°C for 15 min prior to cell inoculation (△).

C. parvum has the ability to self-repair by NER, it is most likely to occur during the early stages of the excystation process at 37°C since DNA repair requires active cellular machinery. In a UV-exposed oocyst that is introduced into a suitable infection host, repair will need to occur before the formation of a trophozoite since cellular differentiation requires active gene expression.

In this study the potential for reactivation was investigated by inoculating UV-exposed oocysts onto HCT-8 cell monolayers in the presence and absence of an array of possible repair conditions. As described above, the infectivity of control and UV-exposed oocysts, with and without exposure to reactivation conditions, was expressed as the logit transformation of proportional infectivity at each oocyst challenge dose (Figure 3.11). The levels of inactivation in the presence and absence of potential repair conditions were determined by comparing the ID_{50} values. All of these potential repair conditions were designed to either stimulate oocysts into a metabolically active state following UV-exposure so that any repair mechanisms would be "switched on" or to provide holding periods to allow any repair mechanisms to function. Results of the reactivation experiments are summarized in Table 3.2 and Figures 3.12 – 3.14). Overall, there was a strong correlation ($r = 0.66$) between the level of inactivation displayed by oocysts incubated under the various repair conditions and oocysts that were inoculated onto cell monolayers immediately following exposure to UV indicating an inability to reactivate and regain pre-exposure levels of infectivity (Figure 3.12). In many instances (18 out of 32 reactivation experiments) there was no or very little difference in the amount of inactivation when comparing oocysts with and without exposure to potential reactivation conditions (Figure 3.13). Oocysts that were exposed to white light at 20°C for 2 h and incubated in 0.75% sodium taurocholate at 37°C for 15 min following UV exposure demonstrated less inactivation compared to oocysts that were not exposed to repair conditions (positive values for inactivation difference in Figure 3.13), indicating possible reactivation. However, the difference in relative infectivity was within the 90% prediction limits for UV inactivation. There were no consistent differences for the rest of the repair conditions.

Table 3.2
Summary of UV reactivation conditions and results

Isolate	Age* (days)	UV type†	Repair condition	UV dose (mJ/cm^2)	Inactivation (Log_{10}):‡ No repair	Repair
Iowa	27	MP	96 h post-inoculation incubation	4.21	1.2	>2.8
				8.01	>1.4	>1.4
Iowa	22	MP	168 h post-inoculation incubation	4.16	>1.1	>1.1
Iowa	23	LP	96 h post-inoculation incubation	0.90	0.7	1.4
Maine	7	MP	96 h post-inoculation incubation	2.21	0.7	0.2
				4.37	1.3	3.0
Maine	14	MP	96 h post-inoculation incubation	2.21	>3.1	2.1
				4.36	>3.1	3.1
				6.37	>3.1	>3.1
Iowa	14	LP	Incubation in 0.75% sodium taurocholate	2.15	1.6	1.9
Iowa	28	MP	Incubation in 0.75% sodium taurocholate	2.16	1.9	1.7
Iowa	23	MP	Incubation at room temp. in dark for 2 h then 0.75% sodium taurocholate	3.22	1.9	1.7
				6.29	2.4	2.5
Iowa	104	MP	Incubation at room temp. in light for 2 h then 0.75% sodium taurocholate	3.35	2.9	>2.3
				6.24	>2.9	>2.5
Iowa	230	MP	Incubation in 0.75% sodium taurocholate	3.95	1.6	>1.9
				6.02	3.0	2.8
Maine	28	MP	Incubation in 0.75% sodium taurocholate	6.40	>2.7	>2.7
KSU-1	38	MP	Incubation at 4°C in dark for 18 h then 0.75% sodium taurocholate	6.33	3.6	3.7
					>3.6§	>3.1§
KSU-1	24	MP	Incubation in 0.75% sodium taurocholate and 1 h at 37°C	3.31	0.4	1.0
				6.24	1.6	1.7
KSU-1	44	MP	Incubation in 10% bleach	4.19	1.2	1.9
Iowa	9	MP	Incubation in 10% bleach	4.37	2.1	1.6
Iowa	14	MP	Incubation at 37°C for 30 min	6.11	2.5	3.4
Iowa	14	MP	Incubation at 4°C for 18 h followed by 15 min at 37°C	6.11	2.5	1.8
Iowa	28	LP	4 h at room temp. in dark	1.80	>1.3	0.7
				3.30	>1.3	>1.3
				5.50	2.0	2.0
Iowa	60	LP	Secondary infection	3.10	1.0	>1.0
				6.00	1.2	2.3
Iowa	22	LP	1 h at room temp. in dark, 30 min at 37°C in dark, 1 h white light, 18 h at 4°C	5.00	2.3	2.3
				5.00	>2.9§	>2.7§

* Number of days post-shedding at the time of UV exposure.
† MP, medium pressure UV; LP, low pressure UV.
‡ The level of UV inactivation was assessed either immediately following exposure to UV light (No repair) or following exposure of the oocysts to the various potential repair conditions (Repair).
§ Inactivation measured using the CD-1 mouse assay.

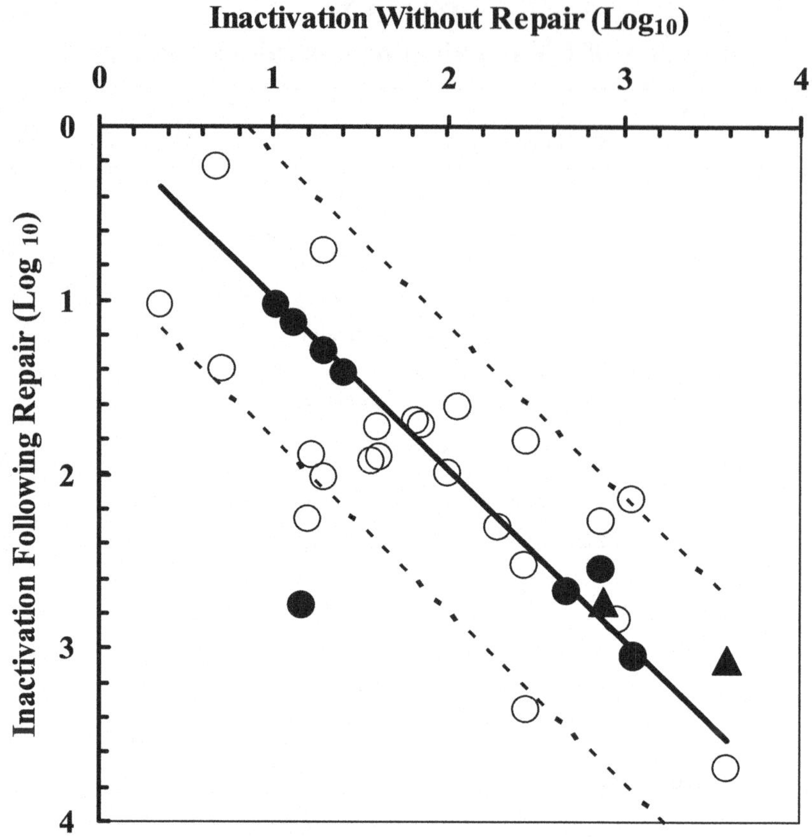

Figure 3.12 Comparison of UV-inactivation of *C. parvum* for oocysts with and without exposures to potential reactivation conditions. Inactivation was assessed by cell culture/RT-PCR (○) or a mouse assay (▲). Open symbols represent actual inactivation data while closed symbols represent minimum levels of inactivation. The dashed lines represent the 90% prediction interval of inactivation of the Iowa isolate, based on data displayed in Figure 3.7.

Cell monolayers were incubated for up to five days following inoculation with UV-exposed oocysts instead of the usual 48 h. UV damaged oocysts may not be able to infect cells immediately following exposure. One mechanism involved in DNA damage repair is cell cycle delay whereby cells temporarily halt most basic cellular metabolism while the damaged DNA is repaired. Therefore, extended incubation on cell monolayers may allow sporozoites time to repair while providing an infection friendly environment. In one instance, increased cell culture incubation time led to an increase in the calculated inactivation but no difference was observed in a second experiment.

Oocysts were also incubated in 0.5% sodium hypochlorite (10% bleach) or 0.75% sodium taurocholate at 37°C for 10 – 20 min prior to inoculating cell monolayers. The latter treatment is known to activate the sporozoites inside oocysts and both are considered excystation triggers. Both of these conditions were also applied after holding periods of up to 18 h under fluorescent white light or in complete darkness at room temperature. Incubation under fluorescent light may allow oocysts to repair if they possess a functional photolyase. Oocysts were also incubated at 37°C for 30 min in PBS prior to inoculating cells. The ID_{50} for oocysts not exposed to UV was 48

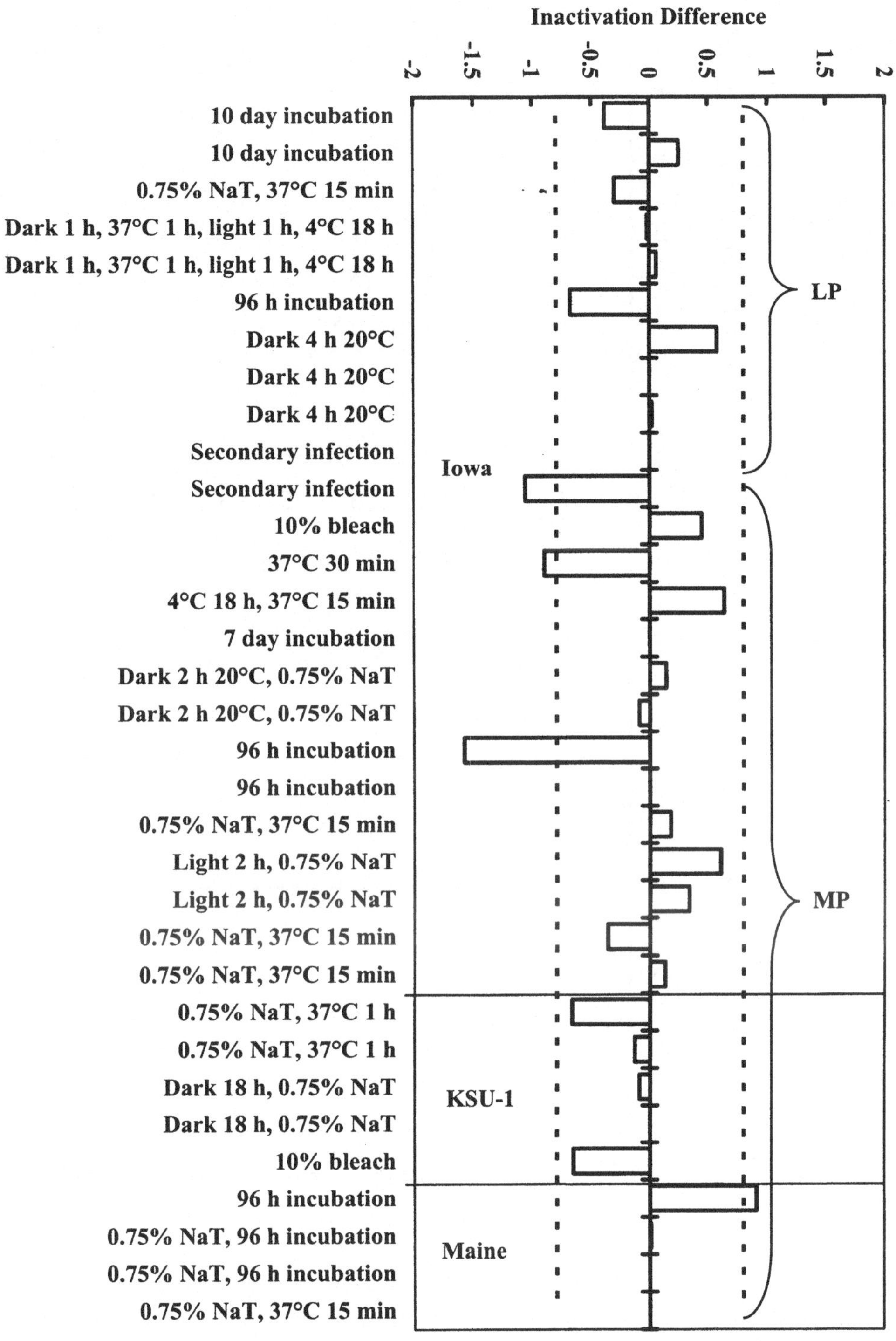

Figure 3.13 Variance in the levels of inactivation for UV-exposed oocysts that were exposed to reactivation conditions (vertical axis) compared to oocysts that were not exposed to reactivation conditions. Oocysts were exposed to varying dosages of medium-pressure (MP) or low-pressure (LP) UV light. Apparent reactivation is represented by positive values on the horizontal axis. Negative values indicate greater inactivation following exposure to repair conditions. Dashed lines represent 90% prediction limits of Iowa isolate inactivation.

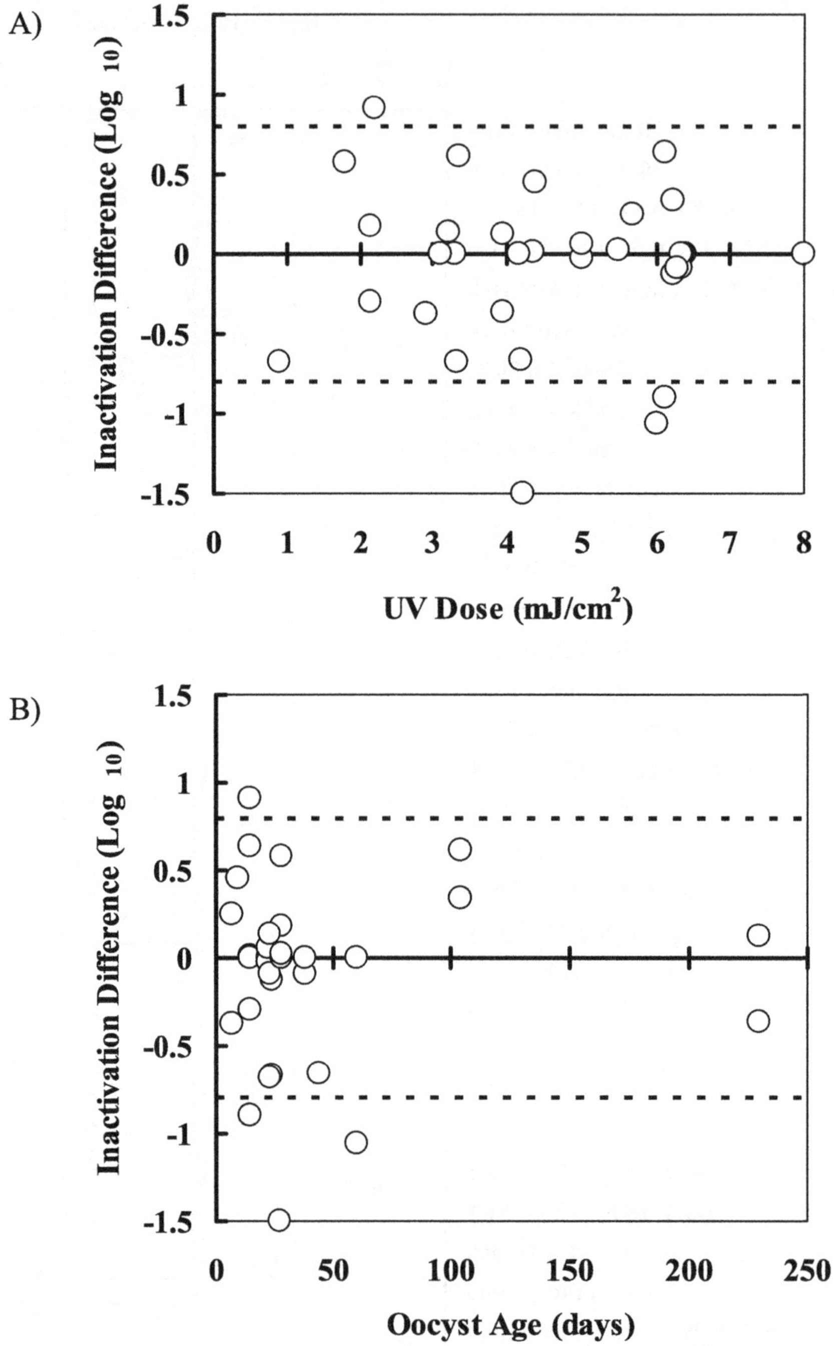

Figure 3.14 Effect of UV dose (A) and oocyst age (B) on the potential for reactivation in UV-treated oocysts that were exposed to various potential reactivation conditions. The dashed lines represent 90% prediction limits of Iowa isolate inactivation. Apparent reactivation is represented by positive values on the vertical axes. Negative values indicate greater inactivation following exposure to repair conditions.

for oocysts that were inoculated onto cells immediately and 97 for oocysts that were incubated at 37°C prior to cell inoculation. There is a relatively high level of variability within the infectivity assay (Rochelle et al., 2002) so these values were not significantly different. The ID_{50} for oocysts exposed to 6.1 mJ/cm^2 of UV light was 13,522 resulting in an inactivation level of 2.5-$\log_{10}$. None of the potential repair conditions generated oocysts with higher levels of infection than the UV-exposed control so inactivation at a dosage of 6.1 mJ/cm^2 was ≥2.2-$\log_{10}$ in all instances regardless of exposure to potential repair conditions in this particular experiment. UV inactivation and reactivation capacity was also compared in fresh and aged oocysts. Medium-pressure UV exposure of 28 day old oocysts (Iowa isolate) resulted in 1.9-$\log_{10}$ inactivation at a UV dosage of 2.2 mJ/cm^2 compared to oocysts that were 230 days post-shedding with 2.3-$\log_{10}$ inactivation. Following incubation in 0.75% sodium taurocholate the inactivation levels were 1.7-$\log_{10}$ and 2.4-$\log_{10}$ for oocysts that were 28 days and 230 days post-shedding, respectively. The inactivation levels for aged oocysts were consistent with the values obtained for fresh oocysts. No reactivation was observed under any of the conditions tested for these older oocysts so aged oocysts appear to respond to UV light and potential repair conditions similarly to fresh oocysts. There was no correlation between oocyst age and the relative levels of inactivation when comparing the inactivation of oocysts exposed to repair conditions and oocysts that were not exposed to repair conditions (Figure 3.14; $r = 0.06$).

The capacity to repair following exposure to UV may be dependent on the magnitude of the UV dose received. An organism may be able to repair the limited number of DNA lesions introduced by low doses of UV but not the increased number generated by higher dosages. Therefore, oocysts were exposed to relatively low dosages of UV (0.9 – 2.2 mJ/cm^2 of low-pressure and medium-pressure UV) and incubated under various repair conditions prior to monolayer inoculation. The inactivation levels for repair exposed oocysts were no lower than those for oocysts not exposed to repair conditions. There was no correlation between UV dose and the capacity to reactivate (Figure 3.14; $r = 0.05$).

In addition, oocysts were held at room temperature in complete darkness for 4 h following UV-exposure and held at 4°C for up to 18 h prior to activation in 0.75% sodium taurocholate. UV-exposed oocysts were also exposed to a series of potential repair conditions by sequential incubation under a combination of conditions. Finally, oocysts were inoculated onto cell monolayers immediately following exposure to UV and incubated as usual. After 48 h incubation at 37°C, the cell monolayer was scraped and homogenized and 10% of the suspension was used to inoculate fresh monolayers to determine if passage through a secondary infection resulted in less inactivation.

In some instances, oocysts exposed to a particular repair condition demonstrated greater inactivation than oocysts that were not exposed to the repair condition (e.g., incubation at 37°C for 30 min; negative values for inactivation difference in Figure 3.13). Overall, the differences in the levels of inactivation for the two types of oocyst (exposed and not exposed to repair conditions) were within the 90% prediction limits of the UV inactivation data for the Iowa isolate. As described above, there was no correlation between the relative levels of inactivation (oocysts exposed to repair conditions versus not exposed to repair conditions) and oocyst age or UV dose. There was also no correlation with UV lamp type (low- or medium-pressure) or the type of assay (cell culture or mice) used to measure infectivity/inactivation.

Since no consistent pattern of reactivation could be discerned within the variability of the assays, the conclusion of these experiments is that *C. parvum* oocysts are not able to reactivate and regain infectivity following exposure to UV light. At least, oocysts cannot reactivate sufficiently to

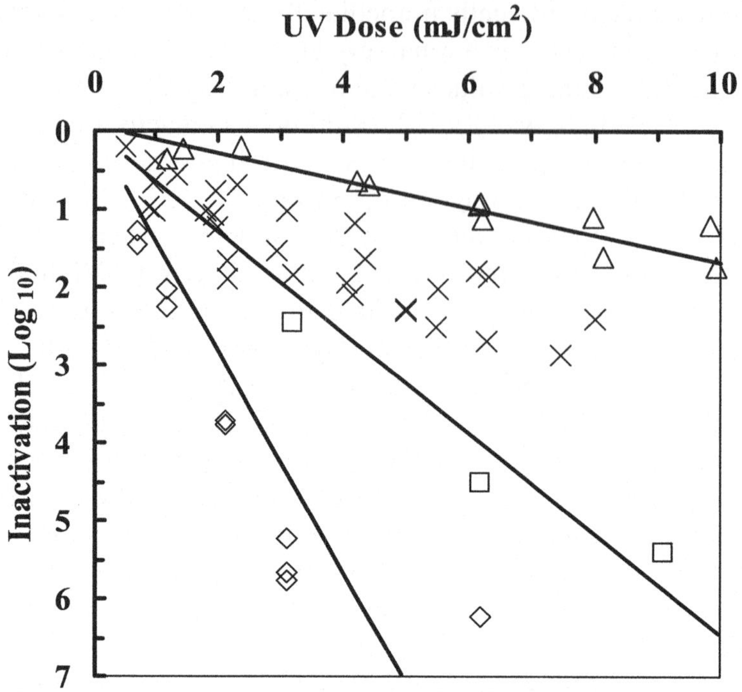

Figure 3.15 Inactivation of *C. parvum* by UV light (×, no line) compared to various strains of repair-deficient (◇ and □) and repair-competent (△) *S. cerevisiae*

regain the levels of infectivity displayed by non-exposed oocysts. While all the evidence to date indicates that oocysts cannot reactivate, the lack of oocyst reactivation in these experiments does not prove that oocysts cannot reactivate following exposure to UV light. It is possible that oocysts can repair and reactivate under a particular combination of conditions that has not yet been evaluated. However, since a variety of potential repair conditions have been investigated in this report and other studies, it seems unlikely that UV-treated oocysts will reactivate under standard drinking water treatment and storage conditions.

All of the studies conducted to date, including this project, have investigated the potential for post-UV reactivation in *C. parvum* by looking for an increase in infectivity following a particular set of repair conditions, relative to UV-irradiated oocysts that were not exposed to these repair conditions. The absence of such an increase in infectivity has been interpreted as an inability to reactivate following exposure to UV light. However, it has been reported that oocysts can repair thymine dimers, based on an endonuclease sensitive site assay (Oguma et al., 2001). So it is possible that *C. parvum* is not only capable of repairing UV-induced damage but is repairing damage in the experiments that have been conducted to date. The level of inactivation displayed by UV-exposed oocysts may represent the maximum level of infectivity of *C. parvum* with all of its repair machinery functioning following exposure to UV. Despite the various efforts to expose irradiated oocysts to potential repair conditions, the very act of inoculating them into a suitable host may provide the necessary repair conditions. Comparison of the UV inactivation dose response curve for *C. parvum* oocysts with the curves for various strains of yeast demonstrated that within the range of 0 – 8 mJ/cm^2, *C. parvum* data was intermediate between repair-competent and repair deficient strains of yeast (Figure 3.15). Since there are no genetically confirmed UV-sensitive mutants of *C. parvum* available for testing, we cannot conclude with absolute certainty that *C. parvum* is repair deficient. The *C. parvum* genome is adenine-thymine (AT)-rich; in protein encoding regions approximately 60% of nucleotides are adenine or thymine, while the non-protein coding regions contain up to 75% AT (Jenkins and Petersen, 1997). Consequently, the *C. parvum* genome may accumulate more UV-induced damage, in the form of thymine dimers, compared to organisms that are guanine-cytosine rich or have a neutral genome composition. The amount of UV-induced damage in such an AT-rich genome may overwhelm any repair that is present, particularly if it is an inefficient mechanism. Pending further investigation, it may be necessary to state that *C. parvum* has an inefficient repair capability rather than no capacity to repair at all. However, from the perspective of the water industry, it can be concluded with reasonable safety that UV-exposed oocysts will not reactivate under post-treatment water storage conditions.

DETECTION OF UV REPAIR GENES

Nucleotide excision repair genes in *C. parvum* were identified primarily by screening the *C. parvum* (Iowa isolate) genomic database using conserved sequences of DNA repair genes from a variety of organisms, including bacteria, *Arabidopsis thaliana*, *Homo sapiens*, and *Saccharomyces cerevisiae*. The genomic database (generated and maintained by the University of Minnesota *Cryptosporidium parvum* Genome sequencing project) consists of a series of contiguous sequences (contigs) covering most of the genome. Each of these contigs contains a fragment of genomic DNA of various sizes and most contain a full gene sequence and flanking DNA sequences. The database contains 2,304 contigs with a total length of 8,262,924 bases (8.3 Mb). The full *C. parvum* genome is estimated to contain 10.4 Mb. Having identified *C. parvum* contigs that contained putative UV repair genes by BLAST screening, the gene was localized within the contig by screening for open reading frames (ORF). The position of the ORF, corresponding to a putative gene within these contigs was bounded by the universal transcription initiation codon (ATG) and one of the stop codons that signal the point on the transcribed mRNA at which protein assembly terminates (TAA, TAG, or TGA; Figure 3.16).

RAD3 and RAD10 were initially found by random sequencing of 200 clones of a genomic DNA library of the KSU-1 isolate (Khramtsov et al. 1995), but the genes were later identified in the *C. parvum* database. Although most open reading frames encoding pertinent genes appeared complete, only partial sequences of three DNA repair genes were found in the database. Thus, the remaining 1,497 nucleotides (43%) of RAD2, 42 nucleotides (7%) of RAD14, and 120 nucleotides (5%) of RAD25 to their respective termination codons were sequenced using the KSU-1 isolate. The full sequence of CpRAD2 was deposited in GenBank with the accession number AY307379. An example of a contig sequence in the *C. parvum* genomic database and the gene it contains is provided in Figure 3.16. Complete genes for nearly all of the major components of the nucleotide excision repair complex were identified in the *C. parvum* genomic database (Table 3.3), with the exception of TFB1. However, a sequence deposited in the GenBank general database (AQ842386) included an ORF that contained a central region displaying significant similarity (when translated to its corresponding protein sequence) with a domain that is conserved in eukaryotic TFB1 proteins (TFIIH 62 kDa subunit; Figure 3.17). Sequence AQ842386 was generated by an earlier *C. parvum* (Iowa isolate) genome sequencing project (Strong and Nelson, 2000). Approximately 48% of this gene sequence (positions 392 – 756) was also present in the *C. parvum* genomic database. Since sequence AQ842386 originated from the Iowa isolate and the MCGP genomic database was also derived from the Iowa isolate, the remaining 52% of the gene was not covered by the sequencing

TTTATCCCCTGCATTTTGTAGACCTAATAGTATGAAAAACATTACGATGTTTCAATCCATTCCACCATGTGTCGAA
TGTATCACAGTACTTATTTCAGATTGCCAAAGTATTTTTAACAATATTTTACCCGATATTAATTCATTTGACCATC
TGCGTGAAGTAAAAGAGAACCTCAGAAAGAAGCAACTTGAAGAGGCTGAAGAAGAGGAAGAAGCAGAAGAAGTAGA
AGGAGAAGAAGAAGAAGAAGAGGAAGAAGAGGAAGAAGGGCAACCAGAATTACTAGGAACCGAGGACAAAGACGAA
GTTGAGAGTGAAGAAAACAACTGAGAACGCTTTTTCTTATTGAAACCAATATCTATTCTTGCCAATTGCTATAGTC
TCATCGCTCGTTTATTCATTAAGGAATTAATGTCAATTTTTTTTATTTTCTTGCAAAGACTATGAAAAAAATTCTC
CCGGCGAGGATCGAACTCGCGACCTTCGGCTCATAAGACCGACGCTCTAACCGAGCTGAGCTACAGGAGAAACTCA
CTGTGTGACGGCTGCAAATTAAAGTTATTCGAATATTGAAGTATCCATTTTGAATAATTTAATTTGTATTAAACTT
TTTTGTAGCCAACAAGTTATTTCTGTAGATCGGCACTTAACATTAGTACCGCTGCCTAAAAAAAAAAAATATTTA
TTAGATAAAATCTGATAGTTATGTACATAAAATCTACTGAATATTAATTTTTGCTTTCATTAATTAGCTTTTAATT
CCCAGTGCACACCTAGTTAGCCTCCGTAGCATAGTGGTAATGCGTTCGCTTCGTAAGCGAAATATCCAGGGTTCAA
TCCCCTGCGTGGGCTGATTTTTGGTAAAATTCGTATTTGAATTAGAATTTTATTTTGTAACTTATTGCCATTGTTT
AATAAGTATACTATTAAAAATTGTACTATGTTAAACCACAGATATTATATTTATTCATATCTCTTGAATATATTTT
TAATGGGGAAAAATGCAGCTATTACCTTTTCAAAAATCAATACTAAATGATATTAATAATCACAAGTCTTGCCTAA
TCATCCTAAGTAAAGGCCTTGGTATTTTTAATATCATCTACAACTACTTGACTAACACAGAAATCAATTCAGGGAA
TTCTATTTTTATTTTAAACTTATCAGTACCTGAGGTTGATTCTTTTAAGCTGTTTGTGCAAAATATGGATCAAAAG
CAACATGATAAAAATAAAGATAAATTTACGAGAAATGGTTCTCTTCTTTCTAGTAAACTTGTGGCTATTAATTCAG
AAATGAATTTAAACAAAAGAAAGGCATTTTATAGTAATGGTGGGATATATATTATCACTTCAAGAATCCTATTAAC
CGATCTATTATCAGAAAAGCTCGATTTTTCACATATAGATGGTATTATATTATTTAATGCAGAGAGCTTAAATGTT
AGAAATTGGAATGATGCGTTTATTCTACAATTATTTAAAAGCAAAAATCCAAATGGGTTCATTAAAGGAATTTCAC
AACGCCCAGAAATTCTAAATCAGGGATATTTTGGGCCCGGAATTACCATGAGATATTTAGGGGCAACTGATTTATT
TTTATACCCAAGAAATAATACTATTGTTGAAGAATCGTTTAAACTTACTCAGAAAATAAAAGTAATAGAAAAGTCA
GTAAAGGCAACTGAATACTTTTATGTTATACAAAAGTGCATTAAAATATTATTGGAAAAAGGGCTACACGAAATTT
TGAAATTTAATCCAAACATTGAATTAACCATGTATGACTTGCTATATAACTCATCAAAAAAATTAAAAAATAAAAT
TGAGTCATTAACGCAAGATTTGTGGTATAAAATGACGTCTAAACACAAACAAATTACAAAGGATATTATATCTTTA
AGAAATTTGCTTGAATTGCTATATTTATCAGATGCATCTGAATTCTTTTTTTGTTTAGAATGCATTCGATCATCAA
AACTTATTGATACTTCTTGGATGTTAACTACAGAATTTGAAACTTTATATAAAGTATCAAGATCAAGAATATTTAG
AATCAACAATGAAGCTAATAATAATAGCAGTGGATCTGAAACTCCTTTTACTCTTTCTCTTGAGGTTAATCCCGTA
TATATTCAATTAATTGATATTATCTCAAATATTGGTAAAGAACTCACTGATGATGATATTTCTACTTTATTACAAG
AAAATGATTTATATGTTAGGCTTTCAGTCAAAAAAAATCAAATTGAATTGAACTCTAACTTAAATTCTTTGGGAAA
AAATGCTGAAACAAAAACTTGTGATGAAACCATTTTTGTAGAAACGGATGGTATCGAATATGGCCTGGATCAAAAC
GAATACTCGCAAATAAATGAAAAATATGTGGAAATATGTTCGAATATATTGAATAATAACTTTGTGCATAAAGTTC
CAGAATATAGAGTTCTACTTATTATTCCTGATGAATTCTACCTTACAATTGCTGAAGTCGAATTTTTGGTACTAAA
TGGCCCACAAAATTTTTCTGCTATGAAATTAATTGAAACTTTACAAAATATCGAATTTGATT***TAGCGTCTGGAGTG***
TCGAATATAATTACGAAAATAATATCTTCTTTTGGGGCTACAACCTGTACTAATTCTAATGAAAAACTTCATTCTT
TTAATCAGCTTAGAGACGTATTATCAAAACTAATTAAAAATAATAGTAATCTAAATTACTCATACTCTTCTAATGA
TTTAATAATTAATTCTATCCCAAATTTGAGTGTATCTGAAAATAATGATTTGAACCAAATAATCAAAAACGACAGA
ATTAATCCAAAAATAATCATTAGTCATCCGAGTGTTAATGTGCAAGGTAAGAAAACAATCGTTTTTCTTCTAAAAT
CTAATTAATTATTAGGGAAATTCTGGAATAT***ACTCTCATTACATTCACCGC***ATCTAATTATAATGTTGGATCCCGA
CATTTCATTGCTCAGGGAAATCGAATTATTTACTGCAACTTCAGAGAAAAGTATAAAATTGTATATACAGGTAATT
ATGATTACAATTCAGAATTCTCTTAG

Figure 3.16 Sequence of Contig1954 from the *Cryptosporidium parvum* genome sequencing project database. The location of start (ATG) and stop codons (TAA) that mark the boundaries of the CpRAD1 gene are indicated along with the PCR primers that were used to detect CpRAD1 in a variety of isolates are also indicated (italics and underlined). The sequence shown is the reverse complement of the sequence retrieved from the database.

approach used for the MCGP database (the database covers approximately 80% of the entire *C. parvum* genome). The gene product of TFB1 is known to be a component of the TFIIH complex and essential for proper UV excision-repair (Wang et al., 1995). Primers based on the database sequence amplified a DNA fragment from the KSU-1 isolate that contained a 594-bp predicted gene product. The full sequence of the CpTFB1 gene was deposited in GenBank with the accession number AY327143.

Based on a recently completed integrated mapping and chromosomal sequencing project of the Iowa isolate of *C. parvum*, genes related to RAD2, RAD4 and RAD10 were located on *C. parvum* chromosome number 6 (Bankier et al., 2003).

Table 3.3
UV-repair genes identified in the *Cryptosporidium parvum* genome (genotype 2)

Gene	Function of protein*	Size (bp)	Database location†
CpRAD1	Endonuclease complex with RAD10, cleaves upstream of DNA damage	1,893	gnl | CVMUMN_5807 | cparvum_Contig1954
CpRAD2	UV-inducible endonuclease, cleaves downstream of DNA damage	3,444	gnl | CVMUMN_5807 | cparvum_Contig990
CpRAD3	Essential component of TFIIH‡ complex, 5'-3' helicase activity	2,523	gnl | CVMUMN_5807 | cparvum_Contig1875
CpRAD4	Interacts with RAD23 and TFIIH	1,524	gnl | CVMUMN_5807 | cparvum_Contig910
CpRAD10	Complexes with RAD1, endonuclease activity cleaving upstream of DNA damage	672	gnl | CVMUMN_5807 | cparvum_Contig1956
CpRAD14	Binds specifically to UV-damaged DNA	630	gnl | CVMUMN_5807 | cparvum_Contig1070
CpRAD23	UV-inducible, attracts TFIIH complex	1,023	gnl | CVMUMN_5807 | cparvum_Contig1934
CpRAD25 (SSL2)	TFIIH component with essential DNA helicase activity	2,412	gnl | CVMUMN_5807 | cparvum_Contig1695
CpRFA1	Subunit of replication factor A	2,064	gnl | CVMUMN_5807 | cparvum_Contig1925
CpSSL1	An essential core component of TFIIH complex	1,128	gnl | CVMUMN_5807 | cparvum_Contig1622
CpTFB1	TFIIH component	594	gnl | CVMUMN_5807 | cparvum_Contig760§
CpTFB2	TFIIH component	573	gnl | CVMUMN_5807 | cparvum_Contig1367
CpTFB3	TFIIH component	771	gnl | CVMUMN_5807 | cparvum_Contig1435

*Function of the protein in *Saccharomyces cerevisiae*.
†Available at www.cbc.umn.edu/ResearchProjects/AGAC/Cp/ (Iowa isolate genomic database).
‡TFIIH, Complex of proteins which acts as a transcription factor.
§Only 48% of this gene was present in the MCPG database.

CTCTATATACGATGATTTGAATCAATTATTACGATATAAGCCTGAATTGGATCAAATTTATAAGAGTTTAGTGCTTA
CAGGAAGTATGTCTTTAGAATCTTTTATTCAAATGCATCATCAAGATATACTTTTATCAAAAACACAAGAAAAAGGA
GTTGAAAATTCTGATTTTTTTTTGAAAAGGCCACCAAAAGTTAGTATCACAAATTCTGGAGG**TATTGATGTAACAAT**
TACTGCCGATGATTTAAAGTCAATTCTTGAAGAAATGCCATCAATTAAAACAAAAATTAAAGAATACGTTCCACATC
GTTTGACAGAAGAAGAATTTTGGAATCGTATTATACAATCACCATTATTTTTTGATATACTTGGGCAAAAAAATACT
CTAGTAAATAATAGTGATACATTAATGATTGGTGAGATTCCTAAAGGAGAAGAATTATTATATGAGTTGAATACACT
TAATAGTAGTAGTAATAATAATGATAATAATGGTAACCAAACAAATGGAATAAGTGAATTACTTGGAAATTATGTTA
GTTCTGAAATCAATCTATTAAATAATGATACTTATAAAAAAATAGGATATGGTACATTAATTGATAATGATTTAAAT
ATTGCATCTGAAAAAAGTGCAAATATTAATACAGGATTCTTTGAAAGATTTAAGGTCATTGAGCTAAATTCTTGAAT
TACAACTGGTATAATGCATATTTATAGACAAATGAATGGATTTGATATTCCAAATTTTTACAACT

Figure 3.17 Sequence of a putative TFB1 gene within a genomic fragment of the Iowa isolate of *C. parvum* (GenBank accession number AQ842386, Strong and Nelson, 2000). The presumed initiation (ATG) and stop (TGA) codons are boxed. The underlined bases indicate a conserved domain in eukaryotic TFB1 subunits (when translated to the corresponding protein sequence).

The MCGP *C. parvum* database was also screened using known sequences from a variety of organisms encoding photolyases (AAA34875, AAB04997, P40115, NP-469940), cyclobutane pyrimidine dimer (CPD)-specific DNA glycosylases (AF378536, AF128168), and UV-specific endonucleases (U70046, AJ243712, NP-005227, NP-015303). Tentative matches were found only in the latter case for one of 15 open reading frames in *Cryptosporidium* contig 1954 and the human and yeast genes encoding UV-specific endonuclease. However, further nucleotide and protein alignments suggested that the open reading frame in *C. parvum* was unrelated to the human or yeast genes. Consequently, no sequences with similarity to photolyases, CPD-specific glycosylases, or UV-specific endonucleases were identified in the *C. parvum* sequence database. UV-specific endonucleases and CPD-specific glycosylases are not widely distributed in nature (McCullough et al., 1998; Yasui and McCready, 1998) and there have been no reports of either type of enzyme in coccidial parasites. Therefore, it is likely that *C. parvum* (and *C. hominis*) does not possess base excision repair (mediated by CPD-specific glycosylases) or UV-specific endonuclease repair mechanisms.

Photoreactivation has been demonstrated in a wide diversity of organisms but there have so far been no reports of photolyases in coccidial parasites. A photolyase gene may not be present in *C. parvum* or, if one is present, it's sequence may be too dissimilar to recognized photolyase genes to be found by similarity searches, or occurs in a region of the genome not yet covered by sequencing projects. Nevertheless, it was reported that UV-exposed oocysts could repair thymine dimers when exposed to fluorescent white light and under dark storage conditions (Oguma et al., 2001), indicating the possible presence of a light-mediated repair mechanism. In a separate study, the same authors also reported that the repair of thymine dimers, assessed by an endonuclease sensitive site assay, was more efficient when oocysts were exposed to fluorescent white light compared to dark repair conditions (Morita et al., 2002). In the current study, oocysts that were exposed to white light for 2 h following UV exposure demonstrated less inactivation (potential photoreactivation) compared to oocysts that were not exposed to repair conditions (Figure 3.13). However, the difference in the level of inactivation was within the 90% prediction limits for UV inactivation so reactivation cannot be concluded in this instance.

Following identification of the 13 NER genes in genotype 2 *C. parvum* (MCGP database) the sequences were compared with the genomic sequence database for a genotype 1 isolate, now referred to as *C. hominis* (VCU/TUSVM genome sequencing project). All of the genes were represented in this second database although most were spread over multiple sequenced fragments (Table 3.4). The level of similarity between the two genomes for the various NER genes ranged from 92 to 98% with an average similarity of 96.4%. In addition the sequences were compared with the general GenBank database to assess the similarity with NER genes in other organisms. For the latter comparison, nucleic acid-based BLAST searches revealed no areas of significant homology. However, performing the searches with translated amino acid sequences revealed significant similarity with a wide variety of non-*Cryptosporidium* organisms demonstrating the high degree of conservation among NER proteins (Table 3.5).

Despite possessing all of the major components of the NER process, our experiments indicated that UV-irradiated *C. parvum* oocysts can not repair sufficiently, if at all, to regain pre-exposure infectivity levels. A possible explanation for this phenomenon is the absence of key regulatory genes that are not intrinsic components of the NER mechanism or the presence of mutations in one or more of the NER genes that reduce or eliminate their activity. Unfortunately, there is currently insufficient information on the sequence elements and motifs within NER genes that are essential for correct function to allow identification of such mutations.

The sequences of CpRAD1, CpRAD2, and CpRAD3 were used to design PCR primers which were then used to amplify DNA extracted from a variety of isolates. Five isolates of *C. parvum* (genotype 2 *C. parvum*) and a single isolate of *C. hominis* (genotype 1 *C. parvum*) all generated amplification products of the expected size (Figure 3.18). The amplicons were sequenced directly and the sequences compared with each other and with the genomic databases. The five *C. parvum* isolates had identical sequences within the region spanned by the PCR primers and were also identical to their corresponding MCGP database sequences (the MCGP database is based on the Iowa isolate). Sequence similarity values for the *C. hominis* sequences compared to the MCGP database were 96% for CpRAD1, 98% for CpRAD2, and 96% for CpRAD3. When compared to the VCU/TUSVM database (*C. hominis*) the sequence similarity values were 100%, 98%, and 100% for CpRAD1, CpRAD2, and CpRAD3, respectively. All of the sequencing that was necessary to complete the identification of UV repair genes was performed on DNA from the KSU-1 isolate as described above. So, these results combined confirm that UV repair genes are widespread in both *C. parvum* (Iowa, KSU-1, TAMU, Maine, Glasgow, and Moredun isolates) and *C. hominis* (NEMC1 isolate and the isolate used for this project).

In this report we have demonstrated that *C. parvum* and *C. hominis* contain nearly all of the genes necessary for nucleotide excision repair, the most common repair mechanism following exposure to UV light. Previous studies have also demonstrated that UV-exposed oocysts are able to repair thymine dimers (Oguma et al., 2001) and that oocysts exposed to UV dosages of up to 123 mJ/cm^2 excyst with the same efficiency as non-exposed oocysts (Bukhari et al., 1999). Thus the parasite should be able to undergo some degree of recovery following UV exposure. However, all of the experiments conducted to date have failed to detect any post-UV parasite development beyond the time of excystation when in-vitro or in-vivo developmental assays are employed. These results imply that the process of UV excision repair may be inefficient in *C. parvum*, being effective only at very low levels of UV damage or the DNA may not be completely repaired, leaving thymine dimers and broken DNA in genes that are essential for parasite replication. Alternatively, there may be no UV repair function at all. If the latter is the case, then the genes encoding UV repair proteins are either not expressed or the products are

Table 3.4
UV-repair genes identified in the *Cryptosporidium hominis* genome (genotype 1 *C. parvum*)

Gene	Size* (bp)	Positions†	Sequence location in database‡	Similarity§
CpRAD1	1,893	1 – 778	CP_A594_H11.G1	761/779 (97%)
		844 – 1416	CP_A297_H12.G1	552/573 (96%)
		1322 – 1893	CP_A324_G07.G1	555/580 (95%)
CpRAD2**	3,444	1 – 213	CP_A342_C11.B1	208/213 (97%)
		302 – 928	CP_A428_D08.B1	609/627 (97%)
		983 – 1616	CP_A806_A12.B1	607/634 (95%)
		1548 – 2275	CP_A593_F12.B1	676/732 (92%)
		2558 – 3348	CP_A598_C03.B1	771/791 (97%)
		3038 – 3444	CP_A417_D06.G1	356/366 (97%)
CpRAD3	2523	1 – 536	CP_A935_D10.B1	526/536 (98%)
		544 – 1337	CP_A094_A09.G1	780/794 (98%)
		1233 – 2042	CP_A308_F08.B1	802/821 (97%)
		1944 – 2523	CP_A308_FO8.G1	700/721 (97%)
CpRAD4	1524	1 – 550	CP_A366_D03.B1	533/550 (96%)
		80 – 847	CP_A600_A12.B1	726/768 (94%)
		808 – 1524	CP_A600_A12.G1	676/718 (94%)
CpRAD10	672	1 – 672	CP_A067_H08.B1	659/672 (98%)
CpRAD14**	630	1 – 134	Not in database	
		135 – 630	CP_A680_B01.G1	481/498 (96%)
CpRAD23	1023	1 – 660	CP_A658_H02.B1	636/664 (96%)
		400 – 1023	CP_A695_E04.B1	611/624 (97%)
CpRAD25**	2412	1 – 622	CP_A638_H10.G1	609/622 (97%)
		597 – 1457	CP_A650_F10.B1	821/864 (95%)
		1049 – 1864	CP_A690_C04.G1	791/818 (96%)
		1650 – 2249	CP_A373_C03.B1	591/600 (98%)
		2250 – 2412	Not in database	
CpRFA1	2064	1 – 785	CP_A140_D10.B1	762/785 (97%)
		600 – 1396	CP_A200_F07.B1	777/801 (97%)
		1312 – 2064	CP_A552_B12.B1	743/754 (98%)
CpSSL1	1128	1 – 570	CP_A300_F07.G1	559/570 (98%)
		402 – 988	CP_A448_G03.B1	567/594 (95%)
		709 – 1120	CP_A898_A11.G1	394/415 (94%)
CpTFB1	603	1 – 245	CP_A476_F08.G1	242/245 (98%)
		6 – 584	CP_A113_D09.B1	574/582 (98%)
CpTFB2	573	1 – 561	CP_A129_F10.G1	549/561 (97%)
		3 – 573	CP_A248_D06.B1	555/571 (97%)
CpTFB3	771	1 – 632	CP_A561_G04.B1	608/632 (96%)
		240 – 771	CP_A809_G12.G1	521/542 (96%)

* Size of gene identified in the Iowa isolate database.
† Most of the identified genes were not represented in the genotype 1 database by a single clone or contig; gene sequences were spread over multiple clones.
‡ *Cryptosporidium hominis* genome (genotype 1 isolate) database number accessible at www.parvum.mic.vcu.edu.
§ Number of identical nucleotides in each overlapping region/total number of nucleotides (%).
** Full gene sequence not represented in database; remaining sequence was obtained from the KSU-1 isolate as part of this project (GenBank accession number for CpRAD2 is AY307379).

Table 3.5
Comparison of *C. parvum* nucleotide excision repair proteins with related proteins*

C. parvum gene	Organism	Protein	Similarity† Identical	Similarity† Positive
CpRAD1	*Saccharomyces cerevisiae*	Rad1	28%	47%
	Schizosaccharomyces pombe	Rad1	22%	43%
CpRAD2	*Plasmodium falciparum*	Rad2	26%	46%
	Drasophila melanogaster	XPG (Rad2 homolog)	24%	40%
CpRAD3	*Arabidopsis thaliana*	XPD (Rad3 homolog)	47%	66%
	Dictyostelium discoideum	RepD (Rad3 homolog)	43%	62%
CpRAD4	*Arabidopsis thaliana*	DNA repair protein	37%	66%
	Caenorhabditis elegans	Rad4	33%	52%
CpRAD10	*Homo sapiens*	ERCC1 (Rad10 homolog)	39%	65%
	Encephalitozoon cuniculi	ERCC1-related	30%	53%
CpRAD14	*Schizosaccharomyces pombe*	Rad14 homolog	31%	49%
	Homo sapiens	XPA (Rad14 homolog)	26%	48%
CpRAD23	*Ictalurus punctatus*	Rad23	35%	54%
	Schizosaccharomyces pombe	Rad23	37%	54%
CpRAD25	*Dictyostelium discoideum*	repB (Rad25 homolog)	46%	64%
	Saccharomyces cerevisiae	Rad25	43%	62%
CpRFA1	*Plasmodium falciparum*	Replication factor A	33%	56%
	Schizosaccharomyces pombe	Replication factor A	27%	48%
CpSSL1	*Saccharomyces cerevisiae*	Ssl1	29%	50%
	Encephalitozoon cuniculi	Ssl1	27%	45%
CpTFB1	*Rattus norvegicus*	62 kDa TFIIH subunit	34%	48%‡
	Saccharomyces cerevisiae	Tfb1	27%	49%§
CpTFB2	*Saccharomyces cerevisiae*	Tfb2	28%	51%
	Homo sapiens	TFIIH subunit	25%	52%
CpTFB3	*Saccharomyces cerevisiae*	Tfb3	25%	44%
	Plasmodium falciparum	Kinase factor (TFIIH)	31%	47%

* The nucleotide sequences of the *C. parvum* RAD genes were compared against the GenBank protein database using the translated BLAST tool.

† Identical similarity refers to exact amino acid matches whereas positive similarity indicates that not all of the amino acids were identical but they were all at least synonymous.

‡ Similarity over central conserved region of 88 amino acids (264 nucleotides).

§ Similarity over region of 65 amino acids (195 nucleotides).

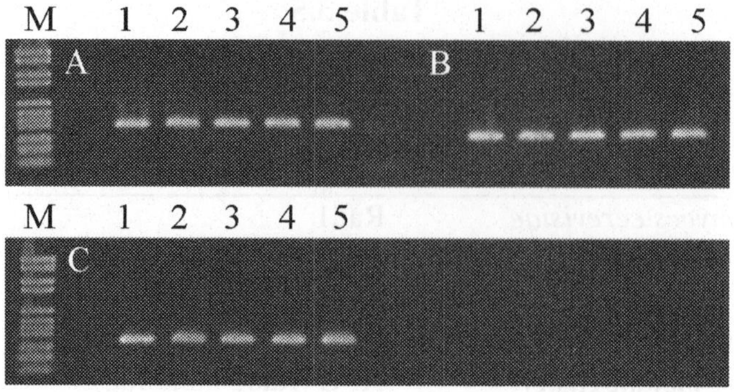

Figure 3.18 Detection of nucleotide excision repair genes in a variety of *C. parvum* isolates. DNA extracted from oocysts of the Iowa (lane 1), TAMU (2), Maine (3), Moredun (4), and Glasgow (5) isolates was amplified using primers targeting the CpRAD1 (Panel A, 369 bp), CpRAD2 (Panel B, 304 bp) and CpRAD3 (Panel C, 345 bp). Following electrophoresis, amplicons were visualized by staining with ethidium bromide and photographing while exposing to UV light using a transilluminator. M, Molecular size standard.

used for non-UV repair functions. Because all of the UV repair gene sequences we identified appeared to be complete and none contained premature termination codons, it is likely that these genes are transcribed and translated and that the proteins have specific function(s) within the parasite. Empirical evidence demonstrated that this was the case for three of the genes. The CpRAD1, CpRAD2, and CpRAD3 primers described above were used in RT-PCR reactions that included RNA extracted from oocyst-inoculated cell monolayers as the template. Control reactions that omitted the reverse transcriptase were negative indicating that the amplification products were generated by mRNA transcribed from the respective genes and not DNA. All cell culture wells that were originally RT-PCR positive using the hsp70 primers (indicating infected cells) were also positive with each of the CpRAD set of primers (Figure 3.19). Cell culture wells that originally negative with the hsp70 primers (no infection) were also negative with the CpRAD primers. These results demonstrate that these three genes are actively transcribed during infection. Amplicons were generated from all three gene transcripts with the same signal strength regardless of whether oocysts were exposed to UV light or post-UV repair conditions. Although many excision repair enzymes have alternate functions within the host cell, some such as RAD4 apparently do not. Further research on protein function is needed to determine the role of these genes in *C. parvum*.

Proteins are also sensitive to UV damage and so, although DNA may be repaired, proteins that are a necessary part of the infection process may be irreversibly damaged. A study of the effect of UV light on proteins demonstrated that microtubule proteins were more sensitive than actin, which was more sensitive than vimentin protein (Rafferty et al., 1997). *C. parvum* contains many microtubules which may play a role in the infectious life cycle.

The RAD1 and RAD2 genes from *C. parvum* KSU-1 were successfully cloned and transformed into RAD1- and RAD2-knockout strains of yeast that were UV-sensitive. Culturing these strains on a defined medium containing galactose should induce expression of the cloned *C. parvum* genes, restoring the DNA repair ability of the yeast mutants. Unfortunately, we were unable to obtain expression of the cloned *C. parvum* genes. The failure to demonstrate cross-species

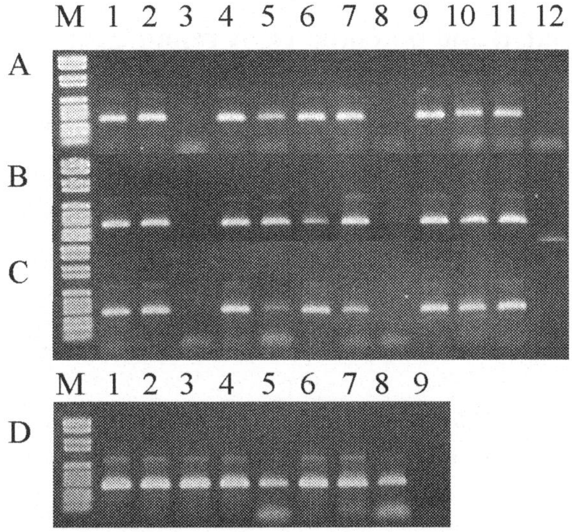

Figure 3.19 Expression of *C. parvum* NER genes during infection of HCT-8 cells. RNA was extracted from cell monolayers infected with the Iowa isolate and amplified by RT-PCR with primers for the CpRAD1 (Panel A), CpRAD2 (Panel B), and CpRAD3 genes (Panel C). Cells were infected with oocysts that had not been exposed to UV (lanes 1 and 2), oocysts exposed to 5 mJ/cm^2 of low-pressure UV (lanes 4 – 7), and UV-exposed oocysts that were incubated in sodium taurocholate to stimulate reactivation (lanes 9 – 11). Lanes 3, 8, and 12 in panels A – C were from uninfected monolayers. In Panel D, hsp70 gene transcripts were amplified from cells infected with oocysts that were not exposed to UV (lanes 1 – 4), exposed to 5 mJ/cm^2 of low-pressure UV (lanes 5 and 6), and incubated in sodium taurocholate following UV exposure (lanes 7 and 8). Lane 9 was from uninfected cells. M, Molecular size standard.

complementation does not necessarily indicate that the *C. parvum* RAD1 and RAD2 gene products are not normally functional in *C. parvum* but may indicate that the proteins do not function in the heterologous system due to inefficient protein-protein interactions with other members of a repair complex (this is particularly an issue with NER proteins).

Additional DNA Repair Genes in *Cryptosporidium*

A variety of additional confirmed and probable DNA repair genes were also identified by further data mining that revealed similarity with known DNA repair genes in other organisms (Table 3.6). The genes that we identified for RAD2 and RAD14 were also represented in the general GenBank database. With the exception of these sequences, the additional repair genes are not involved in repair of UV-induced damage. The functions of the genes that displayed similarity with the *C. parvum* sequences were cell cycle checkpoint control, repair of breaks in double stranded DNA (generally as part of the recombination machinery), mismatch repair, and repair of highly mutagenic modified guanine bases formed by exposure of cells to oxidizing agents and free radicals. Along with the UV-induced DNA damage repair proteins described above, these results demonstrate that *C. parvum* has the genes necessary for a wide array of DNA repair mechanisms.

Table 3.6
Additional RAD-related and potential DNA repair genes identified in *C. parvum*

Accession number*	Similar to:†	Function of protein	Sequence location in database‡
AQ855662	FRP1	Cell cycle checkpoint control	gnl | CVMUMN_5807 | cparvum_Contig867
AQ842602	RAD2§	UV endonuclease	gnl | CVMUMN_5807 | cparvum|_Contig990
AQ411858/ AQ003692	RAD14§	Binds to UV damaged DNA	gnl | CVMUMN_5807 | cparvum_Contig1070
AQ450101/ AQ450107	RAD51**	Maybe involved in repair of double stranded breaks in DNA	gnl | CVMUMN_5807 | cparvum_Contig1652
AQ003650	RAD32**	Repair of double stranded breaks in DNA	gnl | CVMUMN_5807 | cparvum_Contig756
AQ254558	RAD57**	Involved in recombination repair	gnl | CVMUMN_5807 | cparvum_Contig1370
AQ449692	Non PD-DNA glycosylase	Involved in base excision repair	gnl | CVMUMN_5807 | cparvum_Contig1868
AQ856009/ AQ854920	mutS	DNA mismatch repair	gnl | CVMUMN_5807 | cparvum_Contig1883
AQ083791	*A. thaliana* gene	DNA damage repair	gnl | CVMUMN_5807 | cparvum_Contig1825
B88590	mutL	DNA mismatch repair	gnl | CVMUMN_5807 | cparvum_Contig463

* GenBank accession number for *C. parvum* sequences.
† Similarity of *C. parvum* sequence to other identified genes in GenBank.
‡ Available at www.cbc.umn.edu/ResearchProjects/AGAC/Cp/ (Iowa isolate genomic database).
§ The same genes that were identified by screening the Iowa isolate genomic database for this project.
** Although named RAD, these genes may not have an actual DNA repair function.

QUANTIFICATION OF UV-INDUCED DNA DAMAGE

As organisms are exposed to increasing doses of UV light, the genomic DNA within them becomes progressively more damaged accumulating UV-induced photoproducts at an increasing frequency. Since *C. parvum* appears to be sensitive to UV light, it is anticipated that the sporozoite DNA inside oocysts will accumulate increasing amounts of damage with increased UV exposure. The primary type of UV-induced DNA damage is the generation of photoproducts between adjacent bases. The major photoproducts are cyclobutane pyrimidine dimers (CPDs) between adjacent thymine residues (thymine dimers) and a pyrimidine-pyrimidone product (6-4 lesions; Rothschild, 1999). Both types of DNA damage are highly toxic and mutagenic because they prevent passage of both DNA replication and transcriptional machinery. A variety of methods were used to assess the extent of UV-induced DNA damage in oocysts. These methods included quantitative PCR, a Comet assay, an endonuclease site sensitive assay, and an anti-thymine dimer antibody.

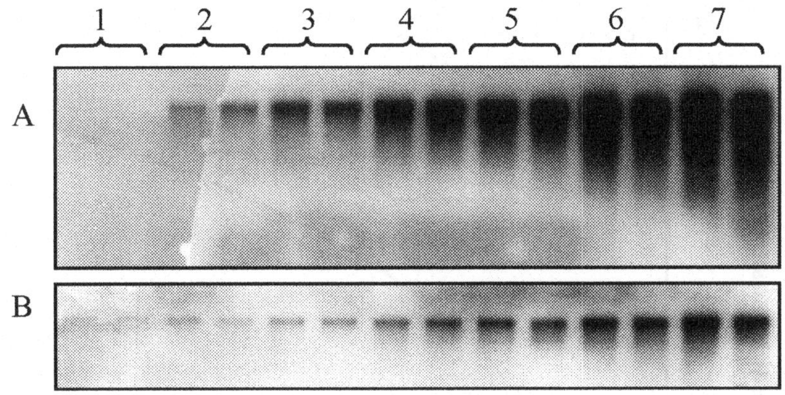

Figure 3.20 Detection of UV-induced thymine dimers using a chemiluminescent antibody assay. Oocysts of the Iowa isolate of *C. parvum* (A) and *S. cerevisiae* cells (B) were exposed to 0 mJ/cm^2 (1), 3.3 mJ/cm^2 (2), 6.8 mJ/cm^2 (3), 12.3 mJ/cm^2 (4), 25.7 mJ/cm^2 (5), 51.6 mJ/cm^2 (6), and 104 mJ/cm^2 (7) of low-pressure UV light. Yeast cells and *C. parvum* oocysts were enumerated using a hemacytometer and diluted so that 3×10^6 cells or oocysts were exposed to UV light in a 10 mL volume.

The Comet assay approach (based on single cell gel electrophoresis) and endonuclease site sensitive assay approach (based on shifting mobility of UV exposed DNA) did not produce useful results. The anti-thymine dimer antibody and quantitative PCR assays are described below.

Detection of Damaged DNA With an Anti-Thymine Dimer Antibody

An assay was developed for directly assessing the amount of UV-induced DNA damage using an antibody specific for cyclobutane thymine dimers. Control experiments demonstrated that the antibody did not detectably bind to DNA extracted from oocysts that had not been exposed to UV light (Figure 3.20, lane 1). Thymine dimers were readily detected in UV-exposed *C. parvum* oocysts following exposure to 3.3 mJ/cm^2 of low pressure UV light (Figure 3.20) and densitometry of the bound antibody signal demonstrated a substantial increase in the accumulation of thymine dimers with increasing UV dosages (Figure 3.21). Thymine dimers were also detectable in DNA extracted from *S. cerevisiae* cells that had been exposed to 3.3 mJ/cm^2 of UV but, in contrast to *C. parvum* oocysts, the accumulation of damage in *S. cerevisiae* was less dramatic. The correlation between UV dosage and the accumulation of thymine dimers was best expressed by a logarithmic relationship for *C. parvum* and a linear relationship for *S. cerevisiae* (Figure 3.21). Although this assay was not tested extensively, the results suggest that it may provide a simple and quantitative method for assessing the level of UV-induced DNA damage in organisms exposed to UV light. The procedure could possibly be used as an alternative to infectivity assays if there is a readily defined relationship between the increase in the number of thymine dimers (quantified using the antibody assay) and the loss of infectivity. The assay may also be used for future studies to determine whether damaged DNA is repaired under certain conditions, even if the oocysts cannot regain pre-UV infectivity levels.

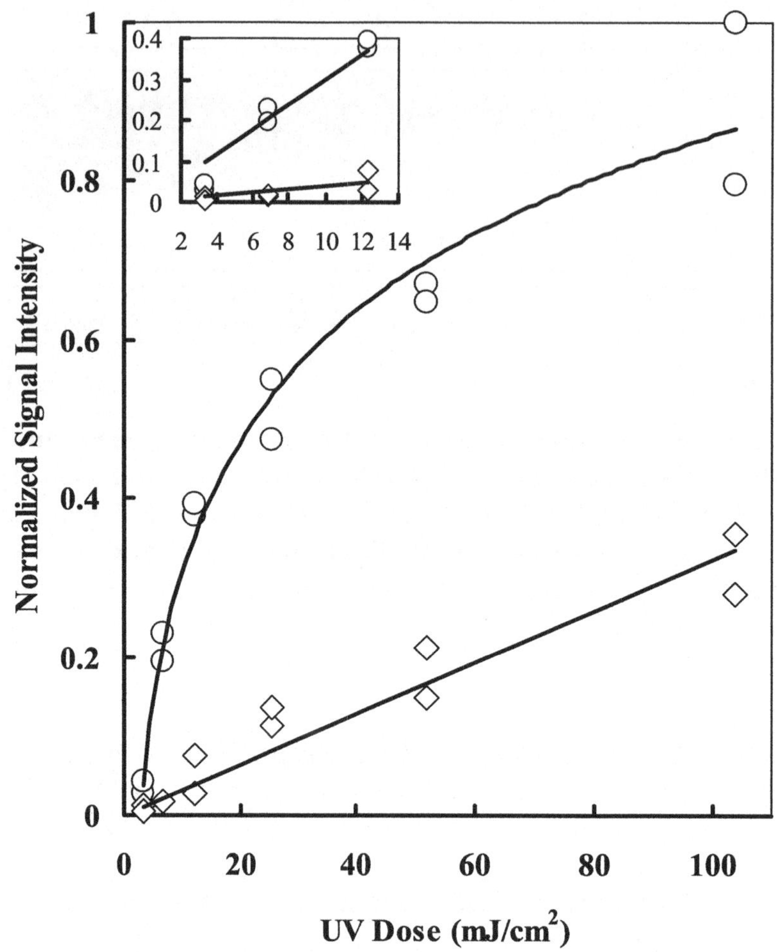

Figure 3.21 Quantification of anti-thymine dimer antibody assay for UV-exposed oocysts of the Iowa isolate of *C. parvum* (○) and *S. cerevisiae* cells (◇). Signal intensity was normalized against the most intense signal (greatest number of thymine dimers). The inset displays just the data for the lower UV dosages.

Detection of Damaged DNA by Quantitative PCR

A further consequence of the DNA damage resulting from exposure to UV light is a reduction or inability of the damaged DNA to serve as a template for DNA polymerase enzymes such as those used in PCR. Therefore, the extent of UV-induced DNA damage was measured using a quantitative PCR method. Quantitative sequence detection (QSD), a real-time quantitative PCR technique targeting the *C. parvum* hsp70 gene was used to directly quantify UV-induced damage to oocyst DNA.

Quantitative PCR methods have been developed for the detection of *C. parvum* oocysts which allow real-time acquisition of amplification results (Di Giovanni et al., 1999). The strategy uses real-time quantitative PCR for the detection of *C. parvum* hsp70 DNA and uses a TaqMan fluorogenic probe and the PE Applied Biosystems 7700 or 7000 Sequence Detector; the approach is called quantitative sequence detection (QSD). The Sequence Detector is a combination of a thermal cycler, laser or fluorometer, CCD camera and data analysis software. QSD is similar to PCR in that it uses DNA (or cDNA produced from mRNA by reverse transcription) as a template and millions of copies of target DNA are synthesized by thermal cycling and the enzyme DNA polymerase. The approach differs from standard PCR in the use of a 5' nuclease assay in the reaction. This method uses the 5' nuclease activity of Taq DNA polymerase to cleave a non-extendable fluorogenic (TaqMan) DNA probe specific for the target DNA sequence being amplified. The fluorogenic DNA probe is cleaved during the extension (synthesis) phase of amplification only if the product being synthesized is the specific DNA target, i.e. any nonspecific amplification products are not detected by the assay, virtually eliminating false positives. The cleavage of the fluorogenic DNA probe releases a fluorescent reporter that is detected by Sequence Detector during the amplification reaction. Immediately after the reaction is complete (approximately 2 to 3 hours), the data is automatically analyzed by the system software and quantitative results, including the specificity of probe confirmation, are available immediately without any manipulation of samples. Our previous research has demonstrated that QSD standard curves targeting the hsp70 gene of *C. parvum* have correlation coefficients greater than 0.95.

Since QSD quantifies the amount of amplifiable DNA in a sample, it can readily be used to measure UV-induced DNA damage and possibly repair. To evaluate DNA repair, oocysts exposed to UV and reactivation conditions were assayed by QSD to determine the extent of original DNA damage and whether any measurable DNA repair has occurred. Such an approach was used to demonstrate repair of UV-damaged DNA in human fibroblasts (Boszko and Rainbow, 1999). While other genetic loci may also be suitable for this type of evaluation, we currently have existing methods for the quantitation of UV-induced damage to the single copy hsp70 gene.

Using a low-pressure UV system (Stratalinker UV crosslinker with integrated calibrated radiometer), oocysts were exposed to 0, 1, 5, 10, 20, 40, 60 100 and 1,000 mJ/cm^2 (Figure 3.22). TaqMan hsp70 QSD data from five sample replicates at each UV dose were compared using analysis of variance (ANOVA). The resolution of the TaqMan hsp70 QSD assay for quantitation of UV-induced damage to *C. parvum* oocysts was determined to be 10 mJ/cm^2 ($P = 0.01$). While mean QSD amplification of oocysts was reduced at the 1 and 5 mJ/cm^2 doses, differences from controls were not statistically significant ($P = 0.10$). QSD amplification at 10, 20, 40, 60 and 100 mJ/cm^2 doses could readily be distinguished from each other ($P \leq 0.01$). Average standard deviation of all replicates for all UV doses was 16.1%. DNA of oocysts exposed to 1,000 mJ/cm^2 was not amplifiable. No PCR inhibitors released from plasticware were detected, since QSD amplification of PCR inhibitor controls (1,000 mJ/cm^2) did not differ significantly from unexposed controls ($P = 0.23$).

Oocysts were also exposed to up to 100 mJ/cm^2 of medium-pressure UV using a Rayox collimated beam unit. ANOVA revealed that a significant difference in TaqMan hsp70 QSD amplification was not observed until 40 mJ/cm^2 ($P = 0.04$; Figure 3.23). QSD data were plotted and the 5 mJ/cm^2 data appear to be outliers, with mean quantitation of 180% of the no exposure control. Despite the variability in the medium-pressure UV data, similar mean reductions in TaqMan hsp70 QSD amplification of oocysts exposed to low-pressure UV at the EP AREC laboratory and medium-pressure UV at MWDSC was observed, particularly at the 20 – 100 mJ/cm^2 doses. A comparison between the low-pressure and medium-pressure induced mean reductions in amplification yield produced a correlation coefficient of 0.96.

A novel QSD assay was also developed to evaluate low dose (<20 mJ/cm^2) UV damage. This assay used randomly amplified polymorphic DNA (RAPD) PCR primers that had previously been used to distinguish different *C. parvum* isolates (Deng and Cliver, 1998). The rationale for this approach was that PCR primers that targeted larger and more numerous genomic regions than

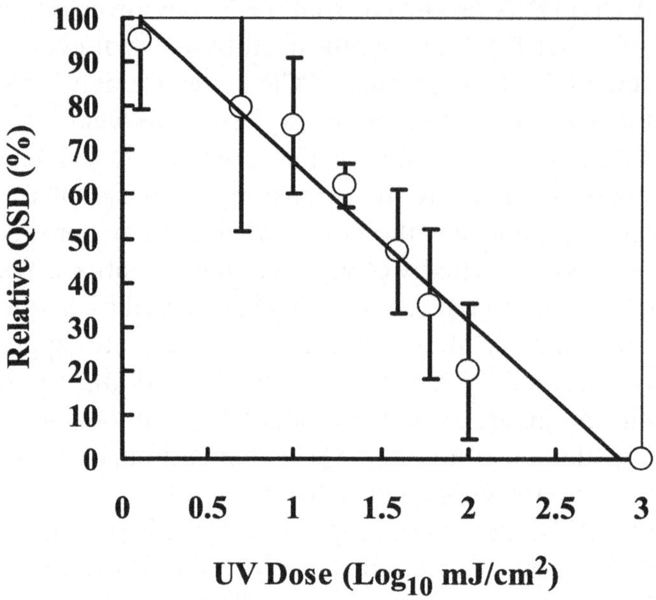

Figure 3.22 TaqMan *hsp70* QSD quantification of low-pressure UV-induced damage to *C. parvum* oocysts. QSD analyses were performed on approximately 200 oocysts. Data are expressed as means and standard deviations of 5 replicates. DNA of oocysts exposed to 1,000 mJ/cm² was not amplifiable.

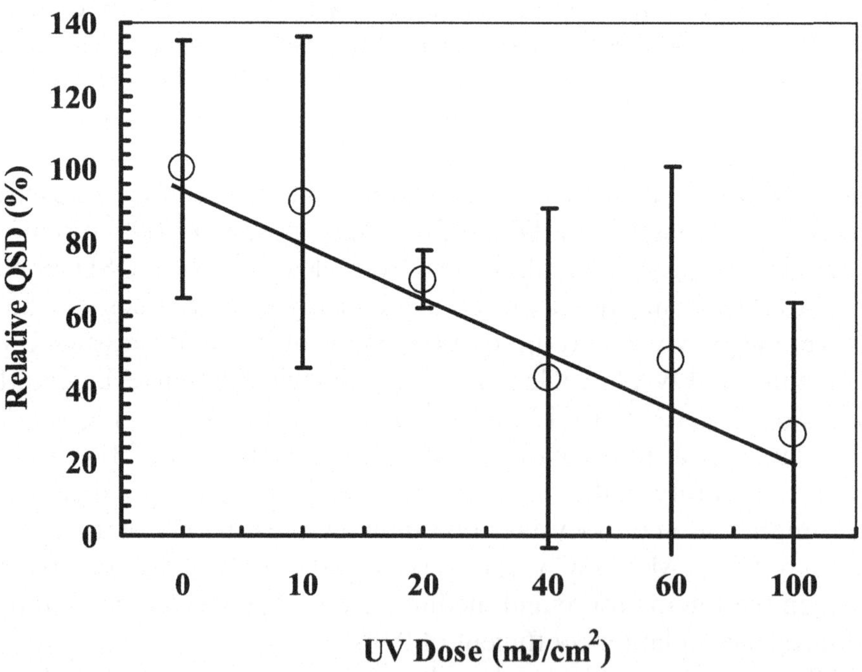

Figure 3.23 TaqMan *hsp70* QSD quantification of medium-pressure UV-induced damage to *C. parvum* oocysts. QSD analyses were performed on approximately 200 oocysts and mean and standard deviations (n = 3) are plotted. 5 mJ/cm² outlier data are excluded.

Table 3.7
Comparison of TaqMan hsp70 and RAPD QSD Ct values for medium-pressure UV exposed oocysts

Medium-pressure UV dose (mJ/cm^2)	QSD Ct value (mean ± standard deviation) for:	
	TaqMan *hsp70* (n = 3)	RAPD (n = 5)
0	33.72 ± 0.42	37.82 ± 0.41
5	32.42 ± 0.41	39.65 ± 4.33
10	33.94 ± 0.67	40.63 ± 6.67
20	34.52 ± 0.02	No Amplification

the hsp70 primers would have a greater chance of detecting lower amounts of UV-induced DNA damage. The same oocyst DNA extracts from the medium-pressure UV experiment described above were evaluated using the RAPD QSD assay. Although there was reduced amplification observed even at the 5 mJ/cm^2 dose, the RAPD QSD cycle threshold (Ct) values of UV-exposed oocysts had large standard deviations (Table 3.7). ANOVAs of RAPD QSD Ct values for 0 and 5 mJ/cm^2 and 0 mJ and 10 mJ/cm^2 treated-oocysts were $P = 0.09$ and $P = 0.04$, respectively. There was no RAPD amplification at the 20 mJ/cm^2 dose.

Similar to the anti-thymine dimer antibody assay, the TaqMan hsp70 QSD method represents a practical surrogate approach for measuring UV-induced damage to *Cryptosporidium* oocysts directly without performing infectivity assays, especially at the 40 – 60 mJ/cm^2 doses likely to be employed for full-scale UV disinfection of drinking water. This method is rapid, providing results from a UV reactor challenge in a matter of hours, and requires relatively low numbers of oocysts (as few as 10^4/L in test waters).

TaqMan hsp70 QSD data for the low-pressure UV treated oocysts were less variable than for the medium-pressure treated oocysts, with significant differences ($P \leq 0.05$) from unexposed controls detected at 10 mJ/cm^2 with low-pressure UV, but not until 40 mJ/cm^2 with medium-pressure UV. The low-pressure UV data had an average standard deviation among replicates of 16.1% (five replicates), while the standard deviation for the medium-pressure data was 36.4% (three replicates). One possible explanation for the variability is that oocysts from the medium-pressure experiments were frozen and thawed one cycle in the absence of Chelex/TE and then shipped by overnight courier between laboratories. This may have had some deleterious effect on the integrity of the DNA. The low-pressure exposures were performed within the same laboratory that performed the QSD analyses. Despite the variability in medium-pressure data, overall reductions in amplification for low- and medium-pressure UV exposed oocysts were similar. Since the low-pressure UV system used for the QSD analyses was equipped with a monochromatic 254 nm light, and the medium-pressure doses were calculated as germicidal doses (based on 260 nm maximum absorbance of DNA), it was expected that they would have similar effects on PCR amplification of oocyst DNA. This is in agreement with the findings of this project (Figure 3.10) and other studies that found no difference in the levels of *C. parvum* inactivation when comparing low-pressure and medium-pressure UV light (This study; Clancy et al., 2000; Craik et al., 2001; Linden and Mofidi, 2003). RAPD QSD results suggested the assay was capable of detecting medium-pressure UV damage to oocysts at doses as low as 5 mJ/cm^2. However, variability in the Ct values prevented the differences from being considered statistically significant ($P = 0.09$). It is important to note that the unexposed oocyst replicates had an expected, low standard deviation of

Ct values (0.41, Table 3.7), similar to the TaqMan hsp70 QSD data (0.42, Table 3.7). Thus, fundamentally the assay was performing as desired. Also noteworthy was the finding that DNA of the 20 mJ/cm^2 exposed oocysts did not amplify with the RAPD primers, whereas complete PCR inhibition was not observed until 1,000 mJ/cm^2 using the TaqMan hsp70 QSD method. Therefore, the RAPD QSD approach appears to be sensitive and shows promise as a tool to assess the level of UV-induced DNA damage, but more development is needed.

Results of this QSD study suggest that careful consideration must be given to the use of PCR for monitoring UV-disinfected drinking water for *Cryptosporidium*. PCR detection of *Cryptosporidium* oocysts was decreased approximately 60% after a UV exposure of 40 – 60 mJ/cm^2, which could result in false negative assays for finished drinking water samples expected to contain low numbers of oocysts. On the other hand, this could be advantageous in that a positive PCR result for a UV-disinfected drinking water sample would indicate that there were very high numbers of oocysts present or that there was ineffective UV disinfection. Future work will investigate the reduction in PCR detection sensitivity for low numbers of oocysts exposed to UV. The QSD method may also be useful for evaluating the UV repair potential of *Cryptosporidium*. To evaluate DNA repair, oocysts exposed to UV may then be subjected to repair conditions such as those described above and assayed by QSD to determine if any measurable DNA repair has occurred. This was attempted for the medium-pressure UV trials but there was no evidence of repair. In fact, the oocysts exposed to potential repair conditions demonstrated less amplification than oocysts not exposed to repair conditions.

CHAPTER 4
SUMMARY AND CONCLUSIONS

The overall conclusion from this investigation is that, although *C. parvum* possesses all of the genes necessary for repair of UV-induced DNA damage by nucleotide excision repair, UV disinfected oocysts will not reactivate under typical treated water storage conditions. Consequently, with respect to *C. parvum*, UV is confirmed as an efficient disinfectant for the water industry. Since typical UV dosages in full-scale UV treatment practice are much higher than the relatively low dosages that result in more than 3-$\log_{10}$ (>99.9%) inactivation, it can be concluded that oocysts in drinking water that are exposed to the full UV dosage will not pose a threat to public health. The specific conclusions of the project are:

1. Oocysts of *Cryptosporidium parvum* were readily inactivated at UV doses well below the 40 mJ/cm^2 typically used operationally in full-scale UV water treatment plants.
2. Inactivation levels at 6 mJ/cm^2 of medium pressure UV light were 2.4-$\log_{10}$ (99.6%) for the Iowa isolate, 2.2-$\log_{10}$ (99.4%) for the Moredun isolate, 2.3-$\log_{10}$ (99.5%) for KSU-1, 1.8-$\log_{10}$ (98.2%) for TAMU, and 2.8-$\log_{10}$ (99.8%) for the Maine isolate.
3. Within the limits of data variability, all isolates appeared to have a similar response to UV light and the inactivation data displayed strong agreement with previous studies.
4. There was no difference in *C. parvum* inactivation between low-pressure and medium-pressure UV lamps.
5. The levels of inactivation displayed by oocysts incubated under various repair conditions were generally no lower than oocysts that were inoculated onto cell monolayers immediately following exposure to UV. Therefore, the conclusion of this study is that *C. parvum* oocysts cannot repair UV-induced damage sufficiently to allow them to regain infectivity following exposure to UV light.
6. All of the major genetic components of the nucleotide excision repair complex, the most common repair mechanism for UV damaged cells, were identified in the genomes of *C. parvum* (previously referred to as zoonotic or genotype 2 *C. parvum*) and *C. hominis* (previously referred to as anthroponotic or genotype 1 *C. parvum*).
7. At least three of the UV repair genes were present in multiple *C. parvum* and *C. hominis* isolates and these genes were expressed during infection of cell cultures.
8. Photolyase, cyclobutane pyrimidine dimers-specific DNA glycosylase, and UV-specific endonuclease genes were not identified in *C. parvum* indicating that nucleotide excision repair may be the only UV repair mechanism available, even though it does not appear to function sufficiently to allow oocyst reactivation.
9. The level of UV-induced DNA damage could be assessed by both molecular (quantitative PCR) and antibody-based methods indicating the potential for measuring the effects of UV light without performing infectivity assays.
10. The consistency of the infectivity and inactivation data generated by this project demonstrates the practical application of cell culture-based infectivity assays within the field of microbiological water quality.
11. The successful identification of UV repair genes in *C. parvum* and *C. hominis* highlights the utility of a genomic/bioinformatic approach, making use of the exponentially

increasing amount of genomic data available in databases, to address microbiological issues within the water industry.

RECOMMENDATIONS

The water industry should consider moving forward with the implementation of UV-based technologies for disinfecting drinking water. Although there are still issues to be resolved, such as the most appropriate UV dosage, configuration of lamps, type and concentration of secondary disinfectants, potential reactivation of UV disinfected *Cryptosporidium* oocysts should not be part of these considerations. Specific recommendations based on the results of this project include:

1. The water industry should not consider repair of UV-induced damage in *C. parvum* when making decisions on the implementation of full-scale UV treatment.
2. Similar repair studies should be conducted with *C. hominis* (genotype 1 *C. parvum*) to ensure that it too cannot regain infectivity following UV disinfection.
3. In-vitro cell culture should be considered as a practical and proven alternative technology for assessing the infectivity and inactivation of *Cryptosporidium*, although the utility of the method with disinfectants other than UV is yet to be demonstrated. Cell culture should also be used to address issues of infectivity, survival, and disinfection of a broader range of pathogens that may be present in water, e.g., *Giardia duodenalis*, microsporidia, and emerging pathogens.
4. The water industry should partner with environmental and regulatory agencies in furthering studies into the genomic/bioinformatic approach for addressing microbiological questions that can not be answered using the traditional "agar plate" method.

FUTURE RESEARCH

All experiments conducted to date have failed to detect any post-UV parasite development beyond the time of excystation when in-vitro or in-vivo developmental assays are employed, although we have demonstrated the presence of NER genes in both *C. parvum* and *C. hominis*. In addition, it has been reported that *C. parvum* can repair thymine dimers. Therefore, UV-irradiated oocysts may be able to repair UV-induced DNA damage but can not regain infectivity. These findings naturally lead to questions such as: Why does the organism carry the genes if it does not or can not use them? and What is missing from *Cryptosporidium*'s repair machinery that prevents reactivation? Even though oocysts carry NER genes, critical regulatory genes may be missing or contain mutations that prevent efficient operation of the repair mechanism. Because all of the UV repair gene sequences we identified appeared to be complete and none contained premature termination codons, it is likely that these genes are transcribed and translated and that the proteins have specific function(s) within the parasite. We demonstrated that three of the genes (CpRAD1, CpRAD2, and CpRAD3) are expressed during infection but other genes may not be expressed. Although many excision repair enzymes have alternate functions within cells, some such as RAD4 apparently do not. Further studies should be conducted to compare the *C. parvum* RAD genes with those in organisms that can reactivate following UV exposure. Also, more detailed studies of gene expression, gene complementation, and protein function will help increase our understanding of the biological basis of DNA repair in *Cryptosporidium* and the lack of reactivation.

Most of the studies conducted to date have looked for evidence of post-UV reactivation in *C. parvum* by looking for an increase in infectivity following a particular set of repair conditions, relative to UV-irradiated oocysts that were not exposed to these repair conditions. The absence of such an increase in infectivity has been interpreted as an inability to repair UV-induced damage. However, it is possible that oocysts may regain infectivity (since they have the necessary genetic components) but the conditions used to study reactivation are restrictive and some vital component that would permit the parasite to regain infectivity is missing from the assays. Therefore, further research should be conducted on developing infectivity assays that optimize the conditions for potential reactivation. It would be beneficial to include human volunteer studies in this work since all of the UV inactivation and repair studies to date have been conducted using surrogate methods, e.g., cell culture or mouse models. The human intestine may provide the ideal environment for post-UV repair, although it is unlikely.

Although we have identified the same suite of UV repair genes in *C. hominis* (previously genotype 1 or anthroponotic *C. parvum*), the two organisms are recognized as separate species with differences in genetics, physiology, and host-range. Therefore, *C. parvum* inactivation and reactivation data can not be safely extrapolated to *C. hominis*. So, to provide an additional measure of safety and confidence in full-scale applications of UV disinfection within the water industry, more detailed studies should be performed to measure the response of *C. hominis* to UV light, the level of inactivation, and the potential for reactivation.

APPENDIX A
GENE SEQUENCES

This appendix contains the full sequences of the nucleotide excision repair genes identified in the genome of *Cryptosporidium parvum*. Each entry starts with the name assigned to the gene (e.g., CpRAD1), followed by the size of the gene and the designation of the genomic sequence in the University of Minnesota *Cryptosporidium parvum* genome sequencing project database (www.cbc.umn.edu/ResearchProjects/AGAC/Cp/). All of the gene sequence information was present within the database with the exception of three genes; the remaining 1497 nucleotides (43%) of CpRAD2, 42 nucleotides (7%) of CpRAD14, and 120 nucleotides (5%) of CpRAD25 were sequenced to their respective termination codons using the KSU-1 isolate of *C. parvum*. Each gene sequence starts with the universal initiation codon (ATG) and ends with one of the three known stop codons (TAA, TAG, and TGA); these codons provide recognition sequences for the beginning and end of genes.

CpRAD1 (1893 bp)   gnl|CVMUMN_5807|cparvum_Contig1954

```
ATGCAGCTATTACCTTTTCAAAAATCAATACTAAATGATATTAATAATCACAAGTCTTGCCTAATCATCCTAAGTAAA
GGCCTTGGTATTTTTAATATCATCTACAACTACTTGACTAACACAGAAATCAATTCAGGGAATTCTATTTTTATTTTA
AACTTATCAGTACCTGAGGTTGATTCTTTTAAGCTGTTTGTGCAAAATATGGATCAAAAGCAACATGATAAAAATAAA
GATAAATTTACGAGAAATGGTTCTCTTCTTTCTAGTAAACTTGTGGCTATTAATTCAGAAATGAATTTAAACAAAAGA
AAGGCATTTTATAGTAATGGTGGGATATATATTATCACTTCAAGAATCCTATTAACCGATCTATTATCAGAAAAGCTC
GATTTTTCACATATAGATGGTATTATATTATTTAATGCAGAGAGCTTAAATGTTAGAAATTGGAATGATGCGTTTATT
CTACAATTATTTAAAAGCAAAAATCCAAATGGGTTCATTAAAGGAATTTCACAACGCCCAGAAATTCTAAATCAGGGA
TATTTTGGGCCCGGAATTACCATGAGATATTTAGGGGCAACTGATTTATTTTTATACCCAAGAAATAATACTATTGTT
GAAGAATCGTTTAAACTTACTCAGAAAATAAAAGTAATAGAAAAGTCAGTAAAGGCAACTGAATACTTTTATGTTATA
CAAAAGTGCATTAAAATATTATTGGAAAAAGGGCTACACGAAATTTTGAAATTTAATCCAAACATTGAATTAACCATG
TATGACTTGCTATATAACTCATCAAAAAAATTAAAAAATAAAATTGAGTCATTAACGCAAGATTTGTGGTATAAAATG
ACGTCTAAACACAAACAAATTACAAAGGATATTATATCTTTAAGAAATTTGCTTGAATTGCTATATTTATCAGATGCA
TCTGAATTCTTTTTTTGTTTAGAATGCATTCGATCATCAAAACTTATTGATACTTCTTGGATGTTAACTACAGAATTT
GAAACTTTATATAAAGTATCAAGATCAAGAATATTTAGAATCAACAATGAAGCTAATAATAATAGCAGTGGATCTGAA
ACTCCTTTTACTCTTTCTCTTGAGGTTAATCCCGTATATATTCAATTAATTGATATTATCTCAAATATTGGTAAAGAA
CTCACTGATGATGATATTTCTACTTTATTACAAGAAAATGATTTATATGTTAGGCTTTCAGTCAAAAAAAATCAAATT
GAATTGAACTCTAACTTAAATTCTTTGGGAAAAAATGCTGAAACAAAAACTTGTGATGAAACCATTTTTGTAGAAACG
GATGGTATCGAATATGGCCTGGATCAAAACGAATACTCGCAAATAAATGAAAAATATGTGGAAATATGTTCGAATATA
TTGAATAATAACTTTGTGCATAAAGTTCCAGAATATAGAGTTCTACTTATTATTCCTGATGAATTCTACCTTACAATT
GCTGAAGTCGAATTTTTGGTACTAAATGGCCCACAAAATTTTTCTGCTATGAAATTAATTGAAACTTTACAAAATATC
GAATTTGATTTAGCGTCTGGAGTGTCGAATATAATTACGAAAATAATATCTTCTTTTGGGGCTACAACCTGTACTAAT
TCTAATGAAAAACTTCATTCTTTTAATCAGCTTAGAGACGTATTATCAAAACTAATTAAAAATAATAGTAATCTAAAT
TACTCATACTCTTCTAATGATTTAATAATTAATTCTATCCCAAATTTGAGTGTATCTGAAAATAATGATTTGAACCAA
ATAATCAAAAACGACAGAATTAATCCAAAAATAATCATTAGTCATCCGAGTGTTAATGTGCAAGGTAAGAAAACAATC
GTTTTTCTTCTAAAATCTAATTAA
```

CpRAD2 (3444 bp) gnl|CVMUMN_5807|cparvum_Contig990
GenBank accession number AY307379

ATGGGGGTTAAAGGTTTGTGGGACATTGTTGCTCCATCTGGAATAAGAGTTAAGCCAGAAACTTTAGAGGGACAAATA
TTGGCTATAGATGCAAGTATTTGGCTAAAGCAATTTTTGATGGGATTAAAGGACAAAGAAGGAAAAATACCTCAAGGT
GCGCATTTATTGGGTTTTTTTAAGAGACTATGTAAACTATTATACTATGGTATATTACCAGTAATAGTTTTTGATGGA
ACTCCTCCAGATATTAAGAAGAGAACTTTGGAGTTAAGAAGAATGCAACGAGAAATGTCAGAAATTAATATGAGAAGA
GTAGCAAATAAATTATTATTAAATACAATAAGACTAAGAGTTGCTAAAAAGGTTAAAGATAAGGTAAGAAAGAGGAAA
GATAAAAAGCTCTCAGATGAAGAATATAAAAACCAAATTGATAATGGTGAGAGTGAGGAAGAAAAAGAAGAAGAACAA
GTACAAGAACAAGAGGAAAGTGAGAGCGAGGAAATTAATCAATCAGCTACAATAAATGATAAAAATAAAGTTGAATAC
AAGAGTAAAAGAAAAGCTCACCATTCAATTCAAGAAAAAATTAGTTTAACTCCCAAATTATTTTCTTCAGAAGATGAG
AATAATATTACTGATGATGATGAAGATGTATATAATAGACCAAATCGTAAATTTAAGAAATATGGAAGATACCAACAA
GTTCATCACTTAATCAATGAAACAGAAGAAAATTCAGAAGCACCAACAAGTTATTTTCCAGTAATGGGATTTTTGAGT
GAGAGAAGAAGATTAGATGAAATGCCTGAAATAGATACAAATTTATTTTCAATGCCAATTCAAATTAGCTCAAGTCAA
AAAATAGCTTCAAAAAAAATGACCAAAATTGAAGATTTAAATGCATTTCTCAGAGCAACTGAATCAAGCAATGATGAA
AGAAGATTAATGGATCTACCATTAGATACAGAAATAGATCCACAAGTATTTGAGCAATTAAACTCAAAACTTCAATAT
GAAATTTTGATCCAACTTAGAGATGCTTGGATTAATGCATTAAGAAGTAATGCTGTTGATACTAAAGATAATATGAGT
CAGTTTAGTAATTCTCAGATTGAGTGTTATTTGAGATATTTGAGAATTAATCAAGAAATAGAGAGATTAAAGGTGAGA
ATGGCTAAAGAATGTGAAGAAGAGCAAATGGAATCAGTTAATGAAGAAATTAAAGTTGTTGAATATGGTCAAATTTAT
GATCCTGACAATAATGATTTTAATAATAATATTCAAGATTGTTCATCTTCTAGTATATTTGAAGATCAAAGTAAAGAA
ACTATCAATGAAGAATTTAATATTGATCTTAATATAGATGATTCTACATTTACTGGTAAAATTATTAACTTTTCAAAG
AAAATTGGTATGGATAATAAAGATATTGATACATTTAATCGTAATGATACATTATTAAGTAGTGATACATTATCAATG
AAAGGATATCATAATTTATCTAAAATTACAAAAAAAGAAAGAATTAAATTTCTTCAAAGTACTGATTACACAAAAAAT
GATAATATTGTATCCAATCCACTAAATACTCCATTATTTGAACTTGAGAAACTTTTGAATAACGAAAATATACAAGAA
AGTCAAATTCAAAATAAAGTAAACAATACCTTTAATGATATTAATGATATTGAAAGACTTGAATTATTAAAAGAAAAT
GTGGATGAATTGTTTAAAATTAATGAACCAATTCAAAAATTTGAAGAAAAAGATGATGGATTTATTGTAGATATTGAG
AATCGAGTTTATCAAAATGAAGATGATAAAGAAGTGAATAAGAAAGATGATTTTGGTAATGAAGGTTTTGATGATAGT
TGTGATGATGATGATGATGATGATTGGCAAGACATTGAATGGGAAAGTGCTAACACAATTCTTACTAAACCACAACAA
AATAATTTTGATTCTTTACAAGTTATTAAAAGTATTATTCTTTCTAATGATTTAGTTTCCAATATTGAGCCAAAACTC
GATAGTATTTTGGGATTAGAAAAGATTGAAGAAAATGATGAATCAAAACAAGAACAAGTTATTTTAATCGACTCTGAG
GAAATTAAAGAACAAAAAAATGATATATATTTGATTGAAGATTCAGGCAAACATGAAGATGGCTTTAAAAAAACGGAA
GTTGGTCAATTCGAAGGTCAAATACAAGATTTGGAACCAGAAAAAGAACAGGAACAAGAGCACCAATCAGAACAGGAA
CCAGAACAAGAACCAGAACAAGAACTAGAACATGAACCTGACCAAGATCAAGATCAAGAACAAAAACCAGAACAAGAA
CCCGAGCGCCAACCAGAACTTGAGCAAGAACCAGAACCTGATCAAAAACAAGAACACCAACCAGAACCCGACCTAGAA
CCAGAACTATTAGACTATTCCTATTTGGAGGATATTAAAGAAATCATCCAAAAGGAAGATGAACAAATCATGGACATG
AACCTTTTTAGTGACCATTATGCAATATATGACAAAAATTTGGATGAAATTCTTTTTGAATTGGAAAAAGAACAAGAA
GAATTAAGCTTAGAATTTACTAAGCATGATATCATTTTGAATTCTGAAATAACAAAAGAAATGCAATATCAAATATGT
CTGCTTTTAAAAGCTCTTGGAATTCCTTGGATTGATTCACCAGGTGAAGCAGAAGCTCAAGCTTCTATTTTAACTCAA
TTAAATATTTGTCATGGTGTATTATCTGATGATTCTGATTGCTTAATTTTCGGAGCAAAAAAAATCTTTCGTAATTTC
TTTTCTGGAAATAGTGTTGAAATGTATGATCTAAATCAAGTTAAAAAGTATCTTGGAATTGAAAAACAAGAACAATTC
TATATTCTAGCAATCTTACTTGGTTGTGATTATACTGTTGGAGTTAATGGAATTGGACCAGTTAATGCTGTTGAAGTA
CTTAAAGCATATCCTGAACTTGAAGATATGATTTTATTCCAAAATTGGTCAATGAATAAGTTTCAAGATAAAAGTAAT
GATCTAATTAATGATACTTTTGAAAGAGCTGAATTTAAAAGAAATCATTCCAATTATAGACACTCTTGGGTATTTCCT
CCAGATTTCCCATGCTTTGATGCAATTCATGCAATGAAAAATCCAAATATTATTTCAAATTTTAAACCAACCTTTGGT
AAAATTGATAAAAATTCAACTATTGAATTAATTACAAATAATACAAATCTAACTCATGAAAAAGTATCTCTAATTATT
GATCCAATCATTAAAAATCAAAATAATAAATACAAACAAACAAAAATTGATGAATTCTTAATTCAAAATCAATCTATT
AAATTTAATCATCATCATATTCAAAAAGATGAAAAAGTTGCATCCATTATTTCTAAAAGAATGAAAAAGGCTCTAAAT
CATCTCAAGTAA

CpRAD3 (2523 bp) gnl|CVMUMN_5807|cparvum_Contig1875

ATGGTTCGGTTTTTTATTGAGGAATTGGAAGTTTTCTTTCCATATGATAATGTCTACCCTGAACAACTTGAGTATATG
AAGTACTTGAAGCAGATTCTAGATGCTCACAGCCATGGAGTTCTTGAAATGCCTACAGGTACTGGTAAAACGGTAACC
TTATTATCTTTTATCACGTCTTATCAGTTGGTTCATCCCAATATGGGTAAGTTAATTTATTGCACTCGTACTGTGGCT
GAAATGGAAAAGGCACTACAGGAGCTTAAAACAGTTGTTGATTATTGTAAGAAGGAAATAGAAAATGATAAGATTAAG
TTAGAGCAAGAAATAAAGAGTGAAAACAATTCCTCTCTTGCGTCAGTCTCTGAGTCTCGTTTTAGTGCTGCTTCCATA
CTGGGAATTGGTATGACTGCTAGAAGAAACATGTGTATCAATCCTAGAGTATCGGTACATGCTGACAGAGACAAGATT
GATTCTATGTGTAGATCAATGACAGCTCCTTGGGTAAGAGCAAAACATCAGATGGAAGCAAGAGAAAGAAGTGCTTTG
AGTGAGGGGGATGCAAATTCTTCAAAAATGACAGAAATCGCCGATATTGAAGAAATGTTGGAGTCTGGATGTACAACT
TTATGTCCATACTATGAAGCTTATGAAAGGGTATGGAGCAGTGATCTTGTTCCTACAGGGATTTACACTATTGATGAA
TTTAAAGACTTTAGTAAGAACTGGGAACATCCAATTTTGGGAAAGAAAATCCAATTTTGCCCATACTTTGCTTCTAAA
AGGCTCATTCAAACTGCTAAAGTTGTCGTACTTAATTACCAATATATTCTAGATCCAAAGGTTGCTCAGGCCTCTCTC
TTAGGAGGTGGAACAGTATCTCAAGGATTCAGCCATGGAGCAAATCCAAATGGACTTAGTCTTGGGTCCAAATTTGCA
GCTTCTCTTCTTCCGGAGCAAGAAGGAACCAAAGAGCCTAGTGTTGTTGTATTTGATGAAGCACACAATATAGATAAC
GTTTGTATTGAAGCTTTATCTGTCAATATGAATAGACAAATCCTAAGCGGCGCTGCACGTAACTTAAGAACTTTGAAA
TCAGAAATCGAAAGTCTCTCTTCATTAGATGAACAGAGGCTACAAGATGAATATACAAGATTAATACAGGGTCTTAGA
AACTCTGGACAAGTTCAGGATGAAGCAGTATTAGAGGATCTTGAAAGGTTTCCTGTACTTCCAGAGGAAATGGAAAAG
ATTAAAAAAGGTTTAATTCCTGGATCAATTAGAAGAGCTGAACACTTTATTACAATTATGAAGAAACTTATTCTGTAT
TTGCAGGAGTATATTCGAGTATATTCAACTAGAATTGAAGGACCTCTGACATTTGTCAAACATATTGAAGCTTCCTGC
TATATTCAGTCTGGACTACTAAAGTTCTGTGATGAAAGACTTAGATCTCTGTTGAATACTTTAAGAATAGTAGACAGT
GATCAATATTCTTCTCTTGAATTAGTTTGTACATTCTTTACAATTTTAGGAAGTTATTCAAAAGGATTTATTGTGATT
GTGGATCCTTATCCTGAGGTTTCTGGACTTTATGACCCAGTAATACAATTAAGCTGTTTAGATTCATCTATTGCTATG
AGGCCAATTTTAAAGCGTTACCAGAGTATTGTTCTAACATCAGGAACTCTTTCTCCTTTGGACCTTTATCCAAAACTA
TTAGGATTTATTCCTGTTGTTAGCCAAAGTCTAACAATGACTTTAGATAGGACATGCATTTGTCCATTGATTGTTACA
AGAGGATCAGATCAGACACCTCTTTCATCAAAATTCGAGTCCAGAGCTGACGTTTCAATTCAACAGAATTATGGTAAA
CTTATTTTAGAAATTACAAAAAAAGTCCCAGATGGGGTAGTTTGCTTCTTTTCTTCATACTTGTATATGGAGCAAATG
CTTTCTCAATGGTATGAATCTGGTTTATTGGCACAAATTATGGAGCATAAGCTAGTATTTGTAGAAACTAAGGATATT
GTATCTACTACTTTAGCTTTACATCATTATAGAAAAGCATGCGATATTGGAAGAGGAGCTATATTCTTTTCTATTGCT
CGTGGAAAAGTTGCTGAGGGAATCGATTTTGATAGACATTATGGGAGGTGTGTCGTAATGGTTGGAATTCCATACCAG
TATACATTATCAAAGATTCTCCAGTCAAGACTTAGCTTTCTAAAGGAGAATTACGGAATTCAGGAAAACGAATTCCTT
ACATTTGATGCAATGAGGCAAGCATCTCAGTGTGTTGGGCGTGTTATTAGGTCAAAGGCAGATTATGGACTTATGATT
TTTGCTGATTTAAGGTATAATAAAAAAGACAAGAGGGAAAAGATTCCTCCTTGGATTTTAAAACATTTAAAGCCTGAA
TATTCAACTTTATCTACGGATATGGCAGTTTCAATTTCTTCAAATTTTTTAAAACAAATGTCACAACCTTATGCTGTT
AGTAACTCAATTATTAAGTTGGGTTCTTAGTAATTTATATAA

CpRAD4 (1524 bp)   gnl|CVMUMN_5807|cparvum_Contig910

ATGCAAAAAGTCAAAAATGAAGCAAGACAGGTCTATATCTTAGGTTGGCTGATCCACTTAAGAATTCTTAATCGAATT
GTTGACAACATTTTCATTAAATCTCACATTTTATCCCTGTACCTATATTATAATCAAGAATATGATACACAAAAAAGT
ATTAGAGGAATATATCAGTGGTTTAGAAAGTATTTCAAAACATATTATTTACCAAATGAATATTGGATTTACAAACTC
CATAATTTTGAAGCGAAGACTAACAATGCGTTTTCGAAATCAAATTGGATGGGTGCATACTATTCTTTATTTTTAAAT
GAAGGTAAAGATTTTGCCCCATGCTTAAATGGAGTAATGTATAATACAATTGGAAGAGTCGAGAATAATTTTCTTGAA
AAATTTCAGGAAATATACAGACAATTCGCACCTCTGATAGATCAACGTAGTTTCTTTCTTGGTAAAATTCTCGGAAGA
ATGATCAGTTCTATTTTATCTGGAAAATGTAGTGCTGAGGCCAGTAACATTCTTTTTGCTGCCCTGGTCCGATCGCTA
GGTCATCATTGCCGCCTAACGTTATTTATTCCTCCTATTATTAACTTTAAAAAAAGTTGTGAATTGCAATGCAACGAT
AGGAGAGCTGAGCTTTGGGTTGAGGTCTTCGAACCAAAGTTCAACAAGTGGATTTCAGTCGATATTCATCGTGGGGG
TGGAACTTCACAGGATGTTCTACAGTTCCGTTTGTAACTTATCCAGAAAAAAATATTTCTTCAATTGAACACAATAGC
GATAAGGGATTGTTTTCAAAACTATTCTTAAGTGAATTTAATGATGACTCTGATGACTCAGACAAAAGCTTAAGTCTT
AAAAAAATAACACCCATATCAGTTATTCCCTTAAAAAAAACTGAAACTAGACAAAGATCAGTCAAATATGAAACGAGT
AATATTGCTGAGCAAGATGAACAAAGAGAGTCCTTAGCATTTGAATTTTCAAATAAAAAGAAAACAGAATCATTAACA
CTCACTCAATTACAGTGCAAAATCGATATCTCAAGGACAAATATTAAACATAAGATTAATGAGAGCATTGGATGGTGG
ATTATATCTGTTAATGAGTATGGGTATTTAAGAGAAACCACTGGAAGATATGTATCAGACTGGTCTCAAGTGTTTCAA
GCTCAACAGAAGAATTCATTTAAACAAAAGGTTGAGTCAATAATATCAACAATAAATTCCACTTCCAATCAGTATAGC
TCTAGGCTGCTTAAATTGGAATTATTAGATGATTTCGAATTGGAAAAAATTATACACGAAAATGATGTAATTCCTATT
TCAAAAACATCTTTCAAAAACCATCCAAAATACGCCATTATTTCATGTCTTAATTCTTTAGAAATCATTCATCCAAAG
GAGCCTATTGTTGGATATTTTCAGGGCGAGCCAATTTATCTTAGAGAAAATGTCCAAACGTTAAAGACTAGGACTCAG
TGGGATCAAGAACAACGCGAAATTAAGATTGGTCAGCAACCATAA

CpRAD10 (672 bp)   gnl|CVMUMN_5807|cparvum_Contig1956

ATGTCAAGCCAAGAAGAAGATATTGTAAAGAAGGAATCCCAGCCTAAGTTTTTTGATGACAAGGCAGGGGAAATGATA
ATAGCCAGTACTAGACAAAGAGGAAATCCAATACTTGCTCATGTGTGTAATGTCCCTTACGATTTTCAGAATATTGTA
CCAGACTTTCTTGTTGGTAAGTATGATGCTGTGGTTTTTATTAGCATAAAATATCACAAGCTTCACAACCAATATTTG
AGGAAGAGGATCGAATCCTTGCAGAAGAACTATAAAGTCAGAATTTTATTATGTCTGGTGGATATTCCTCCTTCTGGT
GCGATTGATGCGGCTATTTTGGAAGTTACTGATATTTGTTTTGACTTAAATATGACTCTGTTTCTTGCTTGGAGTCCA
AAAGAAGCTGGCCATATTTTAGAAACTTTAAAATCTCATGAAAACTCTTCAAGCGAAATCATTAGGGGAGGGCTTTCG
TTAGACTTGTTTTCCAGAATCAGGGACGCCTTGAGCTCCTTACCAAGGATTAACAAGACGGACTCAGAGAATCTTTTA
AAGCATTTTGGATCGATTTCAAAGGTGGTGAATGCTTCCGAAGAAGAGCTTTCAAAGATTCAGGGGATCGGCCCTATA
AAGGCCAAAGTAATTTCGGAAATATTCTCCACAGAGTTCTCAGACTCATAA

CpRAD14 (630 bp)   gnl|CVMUMN_5807|cparvum_Contig1070

ATGTCTCACTCATGTAGGAGTTGTAGATGCGACCTCGGAAATAATCTTTCAGTTGCAGTGGAGGAAACTCTGAAGAGC
GTAGGCATATATGTATGTAGAAGATGCTTTTTTCAAGGGAAAGATGATGAATATAGTAAAATGACTCAGACAAATGCA
ATGAAGGATTATGCATTAAGTGCTCAAGATTTAAGTGGAGAAGGGTTTGCAATAGTCAAGAAACCAACTTCTTATGGG
AAGAACAGATTTATGAAGTTGTACTACAAATTTCAGGTTGAAGAAGCTGCAATAAAAAAATACGGAAGCGTAGAACAA
GCATTGGCTGAGTCAACAAATGCTGAAGATAGAAAAATACAGAAGAAGATAAATAAGAGACTCGGGATAACGAGTGAG
CTAAGCGAAAAGGAACAGAAGAAAATGGAACAAAAAAGGTTAAGGATTGAAAAAAGACTTGAAAATGAAAACAAGAAA
AGGATGTTAAAACAAGAAAGTTACCATGAACATGATTTCGACCCCCCTGTCAAATTAATACAGAAAGTGGTCTTTATA
AGAAAAATGCAAAATCTGTGGATTCGAGCTGACTTGGGAGGAAATTTAGACTCGTTTATATTTGAATTATATACAACT
AAATATTAA

CpRAD23 (1023 bp)   gnl|CVMUMN_5807|cparvum_Contig1934

ATGAAGATTAAGATCAGAACAGTTCAGAATACAGAGATGGAGGTAGAGGTGGAAGCAGATTATTCTATTGAGAAAGTA
AAGCAGGCAATTCAAGAATTGAACCCAGTAATGGAGGCTTCGAGACTAAAGCTCATTTTTGCTGGAAGAATATTGAAT
GATTCTCAAACTGTTCAGGATGTTGGAATAAAAGAGGGAGAGCGATTGGTAGTTCTTTTATCTAAGGGTGCTTCCCAG
AAGGCAGCAGAAAGTCAACAGAATAAACAAAACAATACGTCAAATGAATCAAACACAAATACAGATCCAGCAGCAAGT
GCTACAACTTCAAATATCCAAGCTCAATCAGGGAATTCTGATCCATCAATAGATTCAAGAGCCTCAGCTTTATTAACA
GGGACAGAACTTGAGAAAACTATTACAAATATTGTAAATATGGGTTTCGAAAGAGAACAAGTTACAAGAGCAATGAGA
GCAGCATTTAACAATCCAGATAGAGCTGTTGAGTACCTTACTTCAGGATTACCAATTCCAGAGAATCCAGTGGCACCT
AATCACACTAATATTACACCAGTAAACTCAAATGCTTCTTTGAATGCAGGCTTAACATCATCAGAAGACTTGTCATCA
GAACAGCTCCCGGGAAATTTGGAATCGCTTAGAACTAACCCATTATTCCAACAACTGAGGTCTGTAGTTCAACAAGAT
CCAAGAATCCTCCCAGAGCTACTTGTAAGAATTGGTCAGTCAAATCCTGAGATCCTCCAACTTATTACAGAAAATCAA
GAAGAGTTTATACGAATGATGGAGAGAACTGACTCTGATGAAGTTGGAGAAACTTCCCAATTTCCAATGCAGACAACT
ATTCAGCTTACTCCCCAGGAAGCCGAGTCCGTTGAACGTCTGCAGGCCCTCGGATTCCCAAGAAATGCTGTCATTGAA
GCCTATTTAATTTGTGAAAAGAACGAGGAGCTTGCTGCAAATTATCTTCTCGAAAACAGTGCCGATTTCTTTACTGAT
GGTGCGAACTGA

CpRAD25 (2412 bp)   gnl|CVMUMN_5807|cparvum_Contig1695

ATGGATTTGACAACAAAGTTATCAAGATTTGTGATCAACTATGACCCTAGTTCTAATGATGGAGAGCCCCTCCCTTTG
GCTTGTGACTCTCAACAGGTGGTTTTAAAGCCCTTTAGAGACTATTCTTCAGAATTAAAATTAAAGTCTGATCATGAT
AAAAGACCAATATGGGTTTTTCCAGATGGTTTAATTATCATTGAGACTTTTCATCAAAGTAGTAAAGCTGCATGTGAA
TTTTTGGTGACTATTTCAGAACCTCTTAGTAGACCTGAACTTATACATGAATACCAATTAACCATATTTTCTCTATAT
GCAGCAGTATCACTAGGAATAACAGTTGACTCTATAATAGAAACACTTGGAAAATTTAGTAAGAATGAGATTCCTGAT
ACTTTGGTTAATACAATTAGAGGTCATTGCAAGCTTTTTGGAAAGTTGAAGATTGTTTTATTAGAAGGAAGATACTTT
GTTGAAGCTACAAGTAGACAAGAGTTATCCTATTTACTTTCTGATGAGAGAATTAAGTCTGCTAGAATTATTAGAAAA
GTTAATAATTCTTTAAGTAATTCATCAATTAAAGTATTAAATAACTCAGAATCAAAGCCAGAAGAAGAAGATGCTAAT
AAGTCAGATTTTAATATTACTCAGGCTCCAAGTTTGAATCCAAAAGATCTTGCTTTCAAGATTACTGAGAATCCTAAT
AAAGACAAGGATAATTATGAAAGTTCTGGAAATAGCTTATTACAAACAAACCCAAATTCTAGTTCTAATTCAAACTCG
AATAATATGATTTATTCCTTTGAAATTTCTGGTGATAAGGTTGATATTGTTACAATGGCATCATTTGTTAGTCTTCAT
AGACCTCTGTTATCTGAATATGACTTCAGATCTGACATTAAAAATCCTAATCTAGATATTTCCCTTAAACATACTACT
CAAATTAGATATTATCAAGAACAAGCTTTAAGAATGATGTTTAGTAATGGACGTGCTAGATCTGGAATCATAGTATTA
CCATGTGGAGCAGGAAAAACCTTAACAGGTATTACTGCAGCATGTACTATGAGGAAGTCTGTATTGATATTAACAACA
AGTGCAGTGGCAGTATCACAATGGAAATTTCAATTTGAGCAATATACAACAGTGGATCCAAATAAGGTCCATTGCTTA
ACAAGTAGTAATAAAGAGCCACTTGGCCCTACTTCAGAAGCTGTTGTTCTTGTTACTACGTATACTATGATGGCATTT
ACAGGAAAAAGAAGTGCTTCTTCTTCTCAGATTATTTCACAAATACAAGAAAGAGAATGGGGTTTACTTATATTTGAT
GAAGTTCAGTTTGCACCAGCTCCAGCTTTTAGAAGAATTAATGGAATTGTCAAAGCTCATTGTAAGTTAGGTTTAACA
GCTACACTTGTTAGAGAAGATGATTTGATTCAAGATCTTCAATGGTTAATTGGTCCAAAGTTGTATGAAGCAAATTGG
ATGGAATTGCAAGATAGAGGATATTTAGCTAAAGCTCTTTGTAGTGAAGTATGGTGTCCAATGACAGCAAGCTATTAT
AGAGAGTATCTGGATTGTTCTCATGCAAAGAAAAGAAAACTATGGGTATGTAATCCAAATAAACTAAGAGTTTGTGAG
TTTTTGATACATTGGCATGAACAAAGAGGAGATAAGATTTTAGTTTTCAGTGACTCACTATTTGCATTAATTAATATT
GCAGTAGCTTTGAAGAAACCATTTGTATGTGGTAGTGTTGATACTTTAGAACGAATAAAAATTCTTCAACAATTCAAA
GAAAATCCCAACTTTAATACTATTTTCCTTTCCAAGGTTGGTGATAATGCAATAGATATTCCTCTTGCAAATGTTGTA
ATACAGATATCTTTCAATTTTGCATCAAGACGTCAAGAAGCTCAACGTCTTGGTAGAATTCTAAGACCTAAACCATAT
TCTGGTATGAAATCAGGGAAAGTAATAAACAAGACTTCAGATAGTCCAAATATTTCTGAACAACCATATAATGCTTTC
TTTTATTCATTACTTTCAAAAGATACCGAAGAAATGGAATATGCTGATAAAAGACAGCAATTTATCATTGATCAAGGA
TATAGCTATAGAGTAATTACTATGGACTCTTTTCCATTGAATAACGAGAATTTGCATTATAGTGATATTGAGACACAA
GAAAAGATGCTTGAATTAATTAAACAGAGTGATGATACTTTGGATGATGACGAATTTGGGAATAATATTGATGAAATT
GGAAAGATTTCAAATAATAGTGATATTCTCAGCTCAACAATTACAAAGGTATCTAATATAGATCAGATATCATCAACA
TCATCTGCTTCATTTTCTTTGAATCTATCAAAGAATATAAGTAAAAATCATTCTGTAAATACAAATCAAGCTTGA

CpTFB1 (594 bp)   gnl|CVMUMN_5807|cparvum_Contig760
GenBank accession number AY327143

```
ATGTCTTTAGAATCTTTTATTCAAATGCATCATCAAGATATACTTTTATCAAAAACACAAGAAAAAGGAGTTGAAAAT
TCTGATTTTTTTTTGAAAAGGCCACCAAAAGTTAGTATCACAAATTCTGGAGGTATTGATGTAACAATTACTGCCGAT
GATTTAAAGTCAATTCTTGAAGAAATGCCATCAATTAAAACAAAAATTAAAGAATACGTTCCACATCGTTTGACAGAA
GAAGAATTTTGGAATCGTATTATACAATCACCATTATTTTTTGATATACTTGGGCAAAAAAATACTCTAGTAAATAAT
AGTGATACATTAATGATTGGTGAGATTCCTAAAGGAGAAGAATTATTATATGAGTTGAATACACTTAATAGTAGTAGT
AATAATAATGATAATAATGGTAACCAAACAAATGGAATAAGTGAATTACTTGGAAATTATGTTAGTTCTGAAATCAAT
CTATTAAATAATGATACTTATAAAAAAATAGGATATGGTACATTAATTGATAATGATTTAAATATTGCATCTGAAAAA
AGTGCAAATATTAATACAGGATTCTTTGAAAGATTTAAGGTCATTGAGCTAAATTCTTGA
```

CpTFB2 (573 bp)   gnl|CVMUMN_5807|cparvum_Contig1367

```
ATGAACTCTACTCGTCTCGAGGCAGGAATTATTGTTCAAAGTAACTTTAGGATTTACTGTTATACAGCATCTCCATTA
CAAGCCAAGATTCTTAGACATCTATGTCAAGTTAAGGTTCGTGGCCCAAACATCATTTGTGGAATATTGACTAGGAAA
GGCTTGTTATCCGCATATTCAATGGGAGTGTCAGCTGAACAAATACTTAGATTTTTTTCCAGCAACGCGCATCCAATA
ATATTAAGGCGATTTATGTTAGAAGGAACGAGTATTATACCTGTGAATGTAGAAACGCAGTTGAAACTATGGGAGAAA
GACAAAAAATAGACTCAAAATTTCCCATGCAAGCACCTTTTCAGACTGNGGAGCAAGCCCTAACGATATTCAGCTATTT
AGACAAACGATACTCTATGCAAGAAGTAAAGACATATTACTTTACAATTCCCCCATTGAATTAACAGAAAAAGAACTA
AACTTAAATGTTGAATTACAAAAAAAGATTATATTAGTTATAAAACAGGAATATGAAGATGATATTAAGACTTTTATT
CGAACAAAAAGAGAAGTTATTAATAAATAA
```

CpTFB3 (771 bp)   gnl|CVMUMN_5807|cparvum_Contig1435

```
ATGAATGAAGATATTTGGAAAGATGGTCAAAAACAATATTCTTGCCCATTATGTATGACAAGTACATATTATAAGAAT
GATATCAAAATGTACTATGGTGATCCATGTGGACATAAATTTTGTTCAGAATGTTCAACAAAAGCAAATAATAATAAA
AAATCAACAAGTATATTAAGAGGAAGTAATCAAATATGTCCAGTATGTCATGGTTTTGTTAAATATATTGCAGATTTT
GAATATGGCGAGACAGATTTTTTAAAATTTGAAAGTCAAGCAAGAAAACAGGTATATACAATATTAAATGAAACAAGA
AAAGATTTTAAAGATACTCCTTGTTATGATAATTATCTTGAGAAACGTGAAGATTTAATATATAAATTAATTTATGGA
AATGATACAGAAAAGAAATCAACACAAGATTTCTTAAATCAATATTCAAAAGAAAATCAAGTTGCTATATTAGATAGA
AAAACTAGAGAAGAAACTCTTTTAAGAAATGATATTCTTGAAATTGTACAAAAAGAAAGTACTTTTTATGAGCAACTT
AATCAACTTTCTCAAAATACCACAATTGATCATCTTCAAATTGTACACCCACTTCAAAATGAATATCCTGAATTCTTT
CAAAATTTAATAAAAATCTTACAATCAAAACAATTCTATAATCAAAATATGCTCTTGGTAGTAACATTCCTAATNCTA
TCGACAGGAATATACATGCAAGGATATCTACATCAAATGCTAAGAATCAAAGTTTCACATATTCATCCATGA
```

CpSSL1 (1128 bp) gnl|CVMUMN_5807|cparvum_Contig1622

ATGGATTTTGCAGAGAAATTGGCAGATACTTCAAATAACTCACTGGATGAGCAAGAAGTTGTATATTCATGGGAACAA
GGCCTTATAAAAAATTGGGAGAGATTGGTTGAAACAGAGAGTGGGCTTGTTGTAATTGATGATGAGACTCGCTTTGAT
TCGAATAATGCAGACGAAAGCTATAATATTGAGTTGTTGAAGCGTGGCCCAAATACAAATATTAGGCGTGGTCTACTC
AGAAATATTGTAATTATTTTAGATATGACTTCTAATATGCTGGAGCTTGATTACAAGCCTGACAGACTACAGTGCATG
GTTAAATGCAATGAAATTTTCATTAAACAGCTATTAGAAGATAATCCTCTAACTCAAATCTCGGTTATTTCTATTTAC
GACGGCATTGGCGAAGTAATAATAAGCTACAATTCTAATTTTCTTGAAATTATGACGAGTATTTTGAATTATTTGAAA
AAGGGATGTAAGGGAAGCATGTCTATTCAAAATGGGCTTGAGAAGGCTAAGTATCTACTAGTTTCAATCCCTCCGTAT
GGTACAAAAGAAATTATTTTTTTCTAGGAAGCATGAGAAGTGTAGATAATTCTTTTTTATTCAACGAATGGGTAGAA
GGTTTTTCTTCTAATAATATCATAATTAATGCAATACTCTTCATCCCAGAATTATATATAATTAAGACAATTACAAAG
ATGACAGGGGGTGTTTGTTTATGCGCAATGAATAATGACCACTTACTTAAGTTGACCCTTGAAAATTTCATCAAGCCG
CCTCCAAATAACATTGCGAGCACCCCATTGAACATTAACTTGGTTACAATGGGATTTCCTGAATATGTAAGCAATCAA
ACGCATCCTTTCGGTTGCTCATGTCACAGGTCCTTAACTCATCGTGGGTATTCTTGCCCGAAGTGTAAATCAATTGTT
TGTTACTTACCGACAAAATGTCCGGTTTGCTTGATATATCTTATTTCCCCAAATCACCTTGCAAAATCATTCGCATAT
TTATTTCAGCACCCAGATATTATCTCTATTGAAAATGGGGAATATTTCAAGATTGAGTCCAAGTCTATATACCAAAGT
GCGAATATGCGAGAGCCAGTCTCTAATATGGATATTTGA

CpRFA1 (replication factor A) 2,064 bp gnl|CVMUMN_5807|cparvum_Contig1925

ATGTTTGATAATGGTTCTCCGTTGAGCATAAGGCTGGTAATGTCGGATGCTACTCAGTTACAACCAGGAGATATGATT
AAGATAACAAGATTTTCTTTGAATGAGATCCATTCCACAAAGTTGGTTACTGTTGTTCAGTACGACAAAGTTGGACAT
TGGTCAGGATTTACTTCTATTGGAAAGGCAACTCCACATCCATGTATGGGGTCCAGCAAGGCTGCAGGGGGAGGCCCG
GGGGACCAGAAGCAGCCTGCAGTTCAAAATCAGATGGGAGCTTCATCAGGCACTCCAAATTCAAGTCATTCAACTGAG
CCAAGTAGCGTCAGCATGCAACAACAACAACAGCAACACAATATTCAGATGCAGCCACAACAAGTTGGATACAGAGCG
AGCGGGCCATCAGAACAAATGCCAGGATCCGCAGGACCCATTTCAGGTAGAGATGGCATGCAATCCTCAGGGCCAGGC
TCTATGAATATGGGTAATACAATGAGAAGTGGTCCATACGGAGGAGGTGGAATGGGTTCAGGTCAGTCAAAGGGTCCA
GTTTCAAGGAATCAGGAAATTCCAGTGTATCCAATAAAGAACATAACTTCTTATCTACACAGGTGGAGAATTGTTGGC
CGTGTAGTGAGCAAAAGCGATATCAGAAAGTTCTCATCATCCAAGACAAAGGAAGGCAAAGTGTTTTCATTCGAGATT
TGTGACGCTGATGGGAGCGAGATCAGGGCAACATGTTTCACCAAAGCCGTTGATAAATTCTATGATATGCTGAAAGAG
GGGGAAGTATATAGTTTCTCAAGAGGAGATGTCAAAGAAGCAAATACTCGTTTTAATACTACAGGGCATGAATTTGAA
ATAATCTTCAACGAAGATGCAGAAATTCAAAGTTTGCCACAGGATGATCGGATCCCAAAGAAGACCTTTAATTTTGTT
CCTATTTCAGAAATCCGAGGCTATTCCAAGGGCCAATCACTGGATCTTTTGGGCATTTTACTCAAAGCTGGACCAATT
ACAACAATTACCGTTAAATCTACCGGTGCAGATACCCAGAAAAGAGAACTAACAATCATTGATAAGTCTGGCCACAGC
ATTGACTTGACATTGTGGTCAGAAAGGACCCACTTAGATGAGGGCCTCATTTCTCAAAACCCAGTTATTGCTGTTAAA
AACGCCATTGTTGAGGAGTTTAATGGATTTAGATTGAAACTGGGATCAAGCTCTTCCATTGAATGGAACCCTTCTGGA
GTTGAACAATCGCAGGAACTCAAGAATTGGTTCTTCTCCCAGTCAAATTTACACGAGTCGGTTGTTTCGCTCTCGTCC
TCTAGTCCAAGTCCTGGAAACACTAACTCAAGTTCCTCAGCCAAAAGGCTTAGTATTGATGAAATTGTAGCTACTGCT
ACTTCTGGGTCTAGCTCGTCTGATATGCTCGATGGAGGGATTTGGGTTCACACTTATGGAACTATAAGGTCAATTCGT
GATAATAAGTACTACTGGAGTAGCTGCCCTAAATGTAAGAGAAAAGTTACTGAAATCGAAGATCCTAACAGTATTAAC
GCTTTGGTGTTGCCTTATGGGTCAGAAAAGGAGACTAGTACATCTCTCGGACCAAACTACCACTGTCCAAGTTGCCAG
CAGTCCATTGAAAACCCTATTAAAAAGTACATCTTGAGTTGTGAGATTGTTGACTCTACAGGAACAATCAGAGCCGTA
GCTTTTGCAGAACATGGAGAAACCATAATGGATGGTCTTTCACCAGAAAAACTTGAATCTCTTAAAGAGAATACAGAC
AAAAATACTGAGGATTGCTTCTCTGATAAACTCTTCAGTGAGTGGGTCTTTAAACTGAATGGAAAAAAAGAACTCTAT
CAGGACTCTGAGATAATTAAGTACCGAATCTTTTCAGCTGAAGATATGACAAATCCAGAAATCCTAAACAGAGAGGCC
AAAAGAAAGCTCGAGTATATTTATGACAAGCTTAATAATATTAAGGAAACTAATACTAGTAGCAATGCTTATAGTGCT
GTTAAAGTCGGTATGGGATTTGGACAGTTTGGATATTGA

APPENDIX B
ORGANISMS REFERENCED IN TEXT

Scientific name	Common name
Arabidopsis thaliana	Thale cress (Brassica family of plants)
Caenorhabditis elegans	Soil nematode (approximately 1 mm long)
Cryptosporidium parvum	Previously named *Cryptosporidium parvum* genotype 2
Cryptosporidium hominis	Previously named *Cryptosporidium parvum* genotype 1
Dictyostelium discoideum	Slime mold (Amoeba)
Drasophila melanogaster	Fruit fly
Encephalitozoon cuniculi	A microsporidian pathogen
Homo sapiens	Human
Ictalurus punctatus	Channel catfish
Plasmodium falciparum	Malaria parasite
Rattus norvegicus	Norway rat
Saccharomyces cerevisiae	Baker's yeast (budding yeast)
Schizosaccharomyces pombe	Fission yeast

APPENDIX C
CRYPTOSPORIDIUM GENOME RESOURCES

Database	Institutions	Website	Abbreviation
C. parvum (genotype 2) genome sequencing project (Iowa)	University of Minnesota	www.cbc.umn.edu/ResearchProjects/AGAC/Cp/index.html	MCGP
C. parvum (genotype 1, *C. hominis*) genome sequencing project (NEMC1)	Virginia Commonwealth University and Tufts University School of Veterinary Medicine	www.parvum.mic.vcu.edu/	VCU/TUSVM
C. parvum sequence tags (Iowa)	University of California, San Francisco	medsfgh.ucsf.edu/id/cptags/	UCSF-EST
C. parvum genome HAPPY map	Laboratory of Molecular Biology, Cambridge	www.mrc-lmb.cam.ac.uk/happy/CRYPTO/crypto-genome.html	
Cryptosporidium genome resource	Combined sources	CryptoDB.org	
General sequence database	National Center for Biotechnology Information	www.ncbi.nlm.nih.gov	GenBank

REFERENCES

Arrowood, M.J., and K. Donaldson. 1996. Improved Purification Methods for Calf-Derived *Cryptosporidium parvum* Oocysts Using Discontinuous Sucrose and Cesium Gradients. *Jour. Eukaryot. Microbiol.*, 43:89S.

Atschul, S.F., T.L. Madden, A.A. Schaffer, J. Zhang, Z. Zhang, W. Miller, and D.J. Lipman. 1997. Gapped BLAST and PSI-BLAST: A New Generation of Protein Database Search Programs. *Nucl. Acids Res.*, 25:3389–3402.

Bankier, A.T., H.F. Spriggs, B. Fartmann, B.A. Konfortov, M. Madera, C. Vogel, S.A. Teichmann, A. Ivens, and P.H. Dear. 2003. Integrated Mapping, Chromosomal Sequencing and Sequence Analysis of *Cryptosporidium parvum*. *Genome Res.*, 13:1787–1799.

Blewett, D.A., S.E. Wright, D.P. Casemore, N.E. Booth, and C.E. Jones. 1993. Infective Dose Size Studies on *Cryptosporidium parvum* Using Gnotobiotic Lambs. *Water Sci. Technol.*, 27:61–64.

Boszko, I.P., and A.J. Rainbow. 1999. Host Cell Reactivation of UV-Damaged Reporter Gene in Normal and Xeroderma Pigmentosum Cells Using Quantitative Polymerase Chain Reaction. First Annual Midwest DNA Repair Symposium, Ann Arbor, Mich.

Bukhari, Z., T.M. Hargy, J.R. Bolton, B. Dussert, and J.L. Clancy. 1999. Medium-Pressure UV for Oocyst Inactivation. *Jour. AWWA*, 91(3):86–94.

Bukhari, Z., M.M. Marshall, D.G. Korich, C.R. Fricker, H.V. Smith, J. Rosen, and J.L. Clancy. 2000. Comparison of *Cryptosporidium parvum* Viability and Infectivity Assays Following Ozone Treatment of Oocysts. *Appl. Environ. Microbiol.*, 66:2972–2980.

Campbell, A.T., L.J. Robertson, M.R. Snowball, and H.V. Smith, 1995. Inactivation of Oocysts of *Cryptosporidium parvum* by Ultraviolet Irradiation. *Water Res.*, 29:2583–2586.

Chauret, C.P., C.Z. Radziminski, M. Lepuil, R. Creason, and R.C. Andrews. 2001. Chlorine Dioxide Inactivation of *Cryptosporidium parvum* Oocysts and Bacterial Spore Indicators. *Appl. Environ. Microbiol.*, 67:2993–3001.

Clancy, J.L., Z. Bukhari, T.M. Hargy, J.R. Bolton, B. Dussert, and M.M. Marshall. 2000. Using UV to Inactivate *Cryptosporidium*. *Jour. AWWA*, 92(9):97–104.

Clancy, J.L., T.M. Hargy, D.G. Korich, and M.M. Marshall. 2001. Susceptibility of Multiple Strains of *Cryptosporidium parvum* Oocysts to UV Light [CD-ROM]. In *Proc. of the AWWA Water Quality Technology Conference*. Denver, Colo.: AWWA.

Clancy, J.L., T.M. Hargy, M.M. Marshall, and J.E. Dyksen. 1998. UV Light Inactivation of *Cryptosporidium* Oocysts. *Jour. AWWA*, 90(9):92–102.

Craik, S.A., D. Weldon, and G.R. Finch. 2001. Inactivation of *Cryptosporidium parvum* Oocysts Using Medium- and Low-Pressure Ultraviolet Radiation. *Water Res.*, 35:1387–1398.

Current, W., and N.C. Reese. 1986. A Comparison of Endogenous Development of Three Isolates of *Cryptosporidium* in Suckling Mice. *Jour. Protozool.*, 33:98–108.

Deng, M.Q., and D.O. Cliver. 1998. Differentiation of *Cryptosporidium parvum* Isolates by a Simplified Randomly Amplified Polymorphic DNA Technique. *Appl. Environ. Microbiol.*, 64:1954–1957.

Di Giovanni, G.D., M. Denhart, M.W. LeChevallier, and M. Abbaszadegan. 1999. Quantitation of Intact and Infectious *Cryptosporidium parvum* Oocysts Using Quantitative Sequence Detection (QSD) [CD-ROM]. In *Proc. of the Annual AWWA Water Quality Technology Conference*. Denver, Colo.: AWWA.

Di Giovanni, G.D., F.H. Hashemi, N.J. Shaw, F.A. Abrams, M.W. LeChevallier, and M. Abbaszadegan. 1999. Detection of Infectious *Cryptosporidium parvum* Oocysts in Surface and Filter Backwash Water Samples by Immunomagnetic Separation and Integrated Cell Culture-PCR. *Appl. Environ. Microbiol.*, 65:3427–3432.

Di Giovanni, G.D., M.R. Karim, M.W. LeChevallier, S.N. Boutros, J.S. Chandler, F.A. Abrams, M.L. Spinner, and J.R. Weihe. 2002. Overcoming Molecular Sample Processing Limitations: Quantitative PCR. Vol. 2. Project 00-HHE-2b. Alexandria, Va.: Water Environment Research Foundation and International Water Association.

DuPont, H.L., C.L. Chappell, C.R. Sterling, P.C. Okhuysen, J.B. Rose, and W. Jakubowski. 1995. The Infectivity of *Cryptosporidium parvum* in Healthy Volunteers. *New Engl. Jour. Med.*, 332:855–859.

Fayer, R., U. Morgan, and S.J. Upton. 2000. Epidemiology of *Cryptosporidium*: Transmission, Detection, and Identification. *Int. Jour. Parasitol.*, 30:1305–1322.

Fayer, R., J.M. Trout, L. Xiao, U.M. Morgan, A.A. Lal, and J.P. Dubey. 2001. *Cryptosporidium canis* n. sp. From Domestic Dogs. *Jour. Parasitol.*, 87:1415–1422.

Gatei, W., R.W. Ashford, N.J. Beeching, S.K. Kamwati, J. Greensill, and C.A. Hart. 2002. *Cryptosporidium muris* Infection in an HIV-Infected Adult, Kenya. *Emerg. Infect. Dis.*, 8:204–6.

Guerrant, D.I., S.R. Moore, A.A. Lima, P.D. Patrick, J.B Schorling, and R.L Guerrant. 1999. Association of Early Childhood Diarrhoea and Cryptosporidiosis With Impaired Physical Fitness and Cognitive Function. *Amer. Jour. Trop. Med. Hyg.*, 61:701–713.

Hanes, D.E., R.W. Worobo, P.A. Orlandi, D.H. Burr, M.D. Miliotis, M.G. Robl, J.W. Bier, M.J. Arrowood, J.J. Churey, and G.J. Jackson. 2002. Inactivation of *Cryptosporidium parvum* Oocysts in Fresh Apple Cider by UV Irradiation. *Appl. Environ. Microbiol.*, 68:4168–4172.

Hargy, T.M., J.L. Clancy, Z. Bukhari, and M.M. Marshall. 2000. Shedding Light on the *Cryptosporidium* Threat. *Jour. Environ. Health*, (July/August):19–22.

Hunter, P.R., and Q. Syed. 2001. Community Surveys of Self-Reported Diarrhea Can Dramatically Overestimate the Size of Outbreaks of Waterborne Cryptosporidiosis. *Water Sci. Technol.*, 43:27–30.

Jenkins, M.C., and C. Petersen. 1997. Molecular Biology of *Cryptosporidium.* In Cryptosporidium *and* Cryptosporidiosis. Edited by R. Fayer. Boca Raton, Fla.: CRC Press.

Khramtsov, N.V., M. Tilley, D.S. Blunt, B.A. Montelone, and S.J. Upton. 1995. Cloning and Analysis of *Cryptosporidium parvum* Gene Encoding a Protein With Homology to Cytoplasmic Form hsp70. *Jour. Eukaryot. Microbiol.*, 42:416–422.

Korich, D.G., M.M. Marshall, H.V. Smith, J. O'Grady, Z. Bukhari, C.R., Fricker, J. Rosen, and J.L. Clancy. 2000. Inter-Laboratory Comparison of the CD-1 Neonatal Mouse Logistic Dose-Response Model for *Cryptosporidium parvum* Oocysts. *Jour. Eukaryot. Microbiol.*, 47:294–298.

Korich, D.G., J.R. Mead, M.S. Madore, N.A. Sinclair, and C.R. Sterling. 1990. Effects of Ozone, Chlorine Dioxide, Chlorine, and Monochloramine on *Cryptosporidium parvum* Oocyst Viability. *Appl. Environ. Microbiol.*, 56:1423–1428.

LeChevallier, M.W., and W.D. Norton. 1995. *Giardia* and *Cryptosporidium* in Raw and Finished Water. *Jour. AWWA*, 87:54–68.

Linden, K.G., and A.A. Mofidi. 1999. Measurement of UV Irradiance: Tools and Considerations [CD-ROM]. In *Proc. of the AWWA Water Quality Technology Conference*. Denver, Colo.: AWWA.

Linden, K.G., and A.A. Mofidi. 2003. *Disinfection Efficiency and Dose Measurement of Polychromatic UV Light.* Denver, Colo.: AwwaRF.

MacKenzie, W.R., N.J. Hoxie, M.E. Proctor, M.S. Gradus, K.A. Blair, D.E. Peterson, J.J. Kazmierczak, D.G. Addiss, K.R. Fox, J.B. Rose, and J.P. Davis. 1994. A Massive Outbreak in Milwaukee of *Cryptosporidium* Infection Transmitted Through the Public Water Supply. *New Engl. Jour. Med.*, 331:161–167.

Madore, M.S., J.B. Rose, C.P. Gerba, M.J. Arrowood, and C.R. Sterling. 1987. Occurrence of *Cryptosporidium* Oocysts in Sewage Effluents and Select Surface Waters. *Jour. Parasitol.*, 73:702–705.

McCullough, A.K., M.T. Romberg, S. Nyaga, Y. Wei, T.G. Wood, J.–S. Taylor, J.L. van Etten, M.L. Dodson, and R.S. Lloyd. 1998. Characterization of a Novel *cis-syn* and *trans-syn-II* Pyrimidine Dimer Glycosylase/AP Lyase From a Eukaryotic Algal Virus, *Paramecium bursaria chlorella* Virus-1. *Jour. Biol. Chem.*, 273:13136–13142.

Millard, P.S., K.F. Gensheimer, and D.G. Addiss. 1994. An Outbreak of Cryptosporidiosis From Fresh-Pressed Apple Cider. *Jour. Amer. Med. Assoc.*, 272:1592–1596.

Mofidi, A.A., H. Baribeau, P.A. Rochelle, R. De Leon, B.M. Coffey, and J.F. Green. 2001. Disinfection of *Cryptosporidium parvum* With Polychromatic UV Light. *Jour. AWWA*, 93(6):95–109.

Morgan, U., R. Weber, L. Xiao, I. Sulaiman, R.C. Thompson, W. Ndiritu, A. Lal, A. Moore, and P. Deplazes. 2000. Molecular Characterization of *Cryptosporidium* Isolates Obtained From Human Immunodeficiency Virus-Infected Individuals Living in Switzerland, Kenya, and the United States. *Jour. Clin. Microbiol.*, 38:1180–3.

Morgan-Ryan, U., A. Fall, L.A. Ward, N. Hijjawi, I. Sulaiman, R. Fayer, R.C. Thompson, M. Olson, A. Lal, and L. Xiao. 2002. *Cryptosporidium hominis* n. sp. (Apicomplexa: Cryptosporidiidae) From *Homo sapiens. Jour. Eukaryot. Microbiol.*, 49:433–440.

Morita, S., A. Namikoshi, T. Hirata, K. Oguma, H. Katayama, S. Ohgaki, N. Motoyama, and M. Fujiwara. 2002. Efficacy of UV Irradiation in Inactivating *Cryptosporidium parvum* Oocysts. *Appl. Environ. Microbiol.*, 68:5387–5393.

Morowitz, H.J. 1950. Absorption Effects in Volume Irradiation of Microorganisms. *Science*, 111:229.

National Academy of Sciences. 1996. *Guide for the Care and Use of Laboratory Animals.* Washington, D.C.: National Academy Press.

Nickoloff, J.A., and M.F. Hoekstra. 1998. *DNA Damage and Repair, Volume 1: DNA Repair in Prokaryotes and Lower Eukaryotes.* Totowa, N.J.: Humana Press.

Oguma, K., H. Katayama, H. Mitani, S. Morita, T. Hirata, and S. Ohgaki. 2001. Determination of Pyrimidine Dimers in *Escherichia coli* and *Cryptosporidium parvum* During UV Light Inactivation, Photoreactivation, and Dark Repair. *Appl. Environ. Microbiol.*, 67:4630–4637.

Okhuysen, P.C., C.L. Chappell, J.H. Crabb, C.R, Sterling, and H.L. DuPont. 1999. Virulence of Three Distinct *Cryptosporidium parvum* Isolates for Healthy Adults. *Jour. Infect. Dis.*, 180:1275–1281.

Pedraza-Diaz, S., C. Amar, A.M. Iversen, P.J. Stanley, and J. McLauchlin. 2001. Unusual *Cryptosporidium* Species Recovered From Human Faeces: First Description of *Cryptosporidium felis* and *Cryptosporidium* 'Dog Type' From Patients in England. *Jour. Med. Microbiol.*, 50:293–296.

Pieniazek, N.J., F.J. Bornay-Llinares, S.B. Slemenda, A.J. da Silva, I.N. Moura, M.J. Arrowood, O. Ditrich, and D.G. Addiss. 1999. New *Cryptosporidium* Genotypes in HIV-Infected Persons. *Emerg. Infect. Dis.*, 5:444–449.

Rafferty, N.S., K.A. Rafferty, and S. Zigman. 1997. Comparative Response to UV Irradiation of Cytoskeletal Elements in Rabbit and Skate Lens Epithelial Cells. *Current Eye Res.*, 16:310–319.

Rochelle, P.A., D.M. Ferguson, T.J. Handojo, R. De Leon, M.H. Stewart, and R.L. Wolfe. 1997. An Assay Combining Cell Culture With Reverse Transcriptase-PCR to Detect and Determine Infectivity of Waterborne *Cryptosporidium parvum*. *Appl. Environ. Microbiol.*, 63:2029–2037.

Rochelle, P.A., M.M. Marshall, J.R. Mead, A.M. Johnson, D.G. Korich, J.S. Rosen, and R. De Leon. 2002. Comparison of In-Vitro Cell Culture and a Mouse Assay for Measuring Infectivity of *Cryptosporidium parvum*. *Appl. Environ. Microbiol.*, 68:3809–3817.

Rose, J.B., J.T. Lisle, and M. LeChevallier. 1997. Waterborne Cryptosporidiosis. In Cryptosporidium *and* Cryptosporidiosis. Edited by R. Fayer. Boca Raton, Fla.: CRC Press.

Rothschild, L.J. 1999. The Influence of UV Radiation on Protistan Evolution. *Jour. Eukaryot. Microbiol.*, 46:548–555.

Sambrook, J., P. MacCallum, and D. Russell. 2001. *Molecular Cloning: A Laboratory Manual.* Cold Spring Harbor, N.Y.: Cold Spring Harbor Laboratory Press.

Shaw, J.P., J.P. Malley, and S.A. Willoughby. 2000. Effects of UV Irradiation on Organic Matter. *Jour. AWWA*, 92(4):157–167.

Shin, G.-A., K.G. Linden, M.J. Arrowood, and M.D. Sobsey. 2001. Low-Pressure UV Inactivation and DNA Repair Potential of *Cryptosporidium parvum*. *Appl. Environ. Microbiol.*, 67:3029–3032.

Siede, W. 1998. The Genetics and Biochemistry of the Repair of UV-Induced DNA Damage in *Saccharomyces cerevisiae*. In *DNA Damage and Repair.* Vol. 1. Edited by J.A. Nickoloff and M.F. Hoekstra. Totowa, N.J.: Humana Press.

Strong, W.B., and R.G. Nelson. 2000. Preliminary Profile of the *Cryptosporidium parvum* Genome: An Expressed Sequence Tag and Genome Survey Sequence Analysis. *Mol. Biochem. Parasitol.*, 107:1–32.

Todo, T. 1999. Functional Diversity of the DNA Photolyase/Blue Light Receptor Family. *Mutat. Res.*, 434:89–97.

Tsai, Y.-L., and P.A. Rochelle. 2001. Extraction of Nucleic Acids From Environmental Samples. In *Environmental Molecular Microbiology: Protocols and Applications.* Edited by P.A. Rochelle and P.A. Wymondham. Horizon Scientific Press.

Upton, S.J., M. Tilley, M.V. Nesterenko, and D.B. Brillhart. 1994. A Simple and Reliable Method of Producing In-Vitro Infections of *Cryptosporidium parvum* (Apicomplexa). *FEMS Microbiol. Letts.*, 118:45–50.

Wallis, P.M., S.L. Erlandsen, J.L. Isaac-Renton, M.E. Olson, W.J. Robertson, and H. van Keulen. 1996. Prevalence of *Giardia* Cysts and *Cryptosporidium* Oocysts and Characterization of *Giardia* spp. Cysts Isolated From Drinking Water in Canada. *Appl. Environ. Microbiol.*, 62:2789–2797.

Wang, Z., S. Buratowski, J.Q. Svejstrup, W.J. Feaver, X. Wu, R.D. Kornberg, T.F. Donague, and E.C. Friedberg. 1995. The Yeast TFB1 and SSL1 Genes, Which Encode Subunits of Transcription Factor IIH, Are Required for Nucleotide Excision Repair and RNA Polymerase II Transcription. *Mol. Cell. Biol.*, 15:2288–2293.

Weinert, T., and D. Lydall. 1998. Pathways and Puzzles in DNA Replication and Damage Checkpoints in Yeast. In *DNA Damage and Repair.* Vol. 1. Edited by J.A. Nickoloff and M.F. Hoekstra. Totowa, N.J.: Humana Press.

Xiao, L., C. Bern, J. Limor, I. Sulaiman, J. Roberts, W. Checkley, I. Cabrera, R.H. Gilman, and A.A. Lal. 2001. Identification of 5 Types of *Cryptosporidium* Parasites in Children From Lima, Peru. *Jour. Infect. Dis.*, 185:270–271.

Yasui, A., and A.P.M. Eker. 1998. DNA Photolyases. In *DNA Damage and Repair.* Vol. II. Edited by J.A. Nickoloff and M.F. Hoekstra. Totowa, N.J.: Humana Press.

Yasui, A., and S.J. McCready. 1998. Alternative Repair Pathways for UV-Induced DNA Damage. *BioEssays,* 20:291–297.

Zimmer, J.L., R.M. Slawson, and P.M. Huck. 2001. Inactivation and Potential Reactivation of *Cryptosporidium parvum* Following Medium-Pressure UV Exposure [CD-ROM]. In *Proc. of the Annual AWWA Water Quality Technology Conference*. Denver, Colo.: AWWA.

ABBREVIATIONS

A	adenine
AT	adenine-thymine
ATCC	American Type Culture Collection
AWWA	American Water Works Association
AwwaRF	Awwa Research Foundation
BER	base excision repair
BLAST	basic local alignment similarity
bp	base pairs
BSA	bovine serum albumin
°C	degrees Celsius
C	cytosine
cDNA	complementary DNA
cm	centimeter
cm^2	square centimeter
CO_2	carbon dioxide
CPD	cyclobutane pyrimidine dimer
Ct	cycle threshold for quantitative PCR
CT	product of concentration (mg/L) × time (min)
CTAB	hexadecyltrimethylammonium bromide
d	days
dATP	2'-deoxyadenosine 5'-triphosphate
dCTP	2'-deoxycytidine 5'-triphosphate
DEPC	diethylpyrocarbonate
dGTP	2'-deoxyguanosine 5'-triphosphate
DIG	digoxigenin
DNA	deoxyribonucleic acid
dNTP	deoxynucleotide triphosphate
dTTP	2'-deoxythymidine 5'-triphosphate
dUTP	2'-deoxyuridine 5'-triphosphate
FAD	flavin adenine dinucleotide (oxidized form)
FADH	flavin adenine dinucleotide (reduced form)
FAM	carboxyfluorescein
G	guanine
GAPDH	glyceraldehyde phosphate dehydrogenase

h	hour
HEPES	hydroxyethyl piperazine-1-ethanesulfonic acid
hsp70	heat shock protein (70 kD)
ID_{50}	50% infectious dose
kcal/day	kilocalories per day
KCl	potassium chloride
kD	kilo dalton
KSU	Kansas State University
L	liter
L/min	liters per minute
LP	low pressure
M	molar
MCPG	Minnesota *Cryptosporidium parvum* Genome
$MgCl_2$	magnesium chloride
mgd	million gallons per day
min	minute
mJ/cm^2	millijoules per square centimeter
mM	millimolar
MP	medium-pressure
mRNA	messenger ribonucleic acid
MuLV	murine leukemia virus
MWD	Metropolitan Water District of Southern California
µg	microgram
µL	microliter
µM	micromolar
NaOH	sodium hydroxide
NER	nucleotide excision repair
nm	nanometer
ORF	open reading frame
P	proportional infectivity
PBS	phosphate-buffered saline
PCR	polymerase chain reaction
PD	pyrimidine dimer
QC	quality control
QPCR	quantitative polymerase chain reaction
QSD	quantitative sequence detection

RAPD	random amplified polymorphic DNA
RL	response logit
RNA	ribonucleic acid
RPMI	Roswell Park Memorial Institute
RT	reverse transcriptase
RT-PCR	reverse transcription-polymerase chain reaction
T	thymine
TAMRA	tetramethylrhodamine
TE	tris-ethylenediaminetetraacetic acid
U	unit
UCSF-EST	University of California, San Francisco–Expressed Sequence Tags
UDG	uracil-DNA glycosylase
U/L	units per liter
UV	ultraviolet
VCU/TUSVM	Virginia Commonwealth University/Tufts University School of Veterinary Medicine

IWA Publishing's authorised EU representative for General Product Safety Regulations is Diane D'Arras, 15 rue Duret, 75116 Paris, France, e-mail: safety@iwap.co.uk.

Printed and bound by CPI Group (UK) Ltd, Croydon, CR0 4YY
17/09/2026
02218793-0015